Spatial use and habitat selection of white-tailed eagles (*Haliaeetus albicilla*) in Germany

Dissertation zur Erlangung des akademischen Grades des
Doktors der Naturwissenschaften (Dr. rer. nat.)

Eingereicht im Fachbereich Biologie, Chemie, Pharmazie
der Freien Universität Berlin

vorgelegt von
Friederike Scholz
aus Wurzen

Berlin, Juli 2010

Bibliografische Information der Deutschen Nationalbibliothek

Die Deutsche Nationalbibliothek verzeichnet diese Publikation in der Deutschen Nationalbibliografie; detaillierte bibliografische Daten sind im Internet über http://dnb.d-nb.de abrufbar.

ISBN 978-3-8325-2700-6

Logos Verlag Berlin GmbH
Comeniushof, Gubener Str. 47,
10243 Berlin
Tel.: +49 (0)30 42 85 10 90
Fax: +49 (0)30 42 85 10 92
INTERNET: http://www.logos-verlag.de

Diese Dissertation wurde am Leibniz-Institut für Zoo- und Wildtierforschung Berlin (Direktor: Prof. Dr. Heribert Hofer) im Zeitraum Juli 2006 bis Juli 2010 angefertigt und am Institut für Biologie der Freien Universität Berlin eingereicht.

1. Gutachter: Prof. Dr. Heribert Hofer
2. Gutachter: Prof. Dr. Silke Kipper

Disputation am 8. September 2010

This thesis is based on the following manuscripts:

1. Scholz F and Krone O (submitted). Roaming through the neighbourhood: ranging behaviour and extraterritorial movements of white-tailed eagles (*Haliaeetus albicilla*)

2. Scholz F, Hofer H, and Krone O (submitted). Raptors in highly altered landscapes: Patterns of habitat selection and their implications for the conservation of the recovering white-tailed eagle (*Haliaeetus albicilla*) population in Germany

3. Scholz F, Sulawa J, and Krone O (submitted). Predicting suitable sites for recolonisation by a recovering and expanding raptor population: white-tailed eagles in a central European landscape

CONTENT

CHAPTER 1

General introduction and outline of thesis

Global threats to biodiversity and the wildlife-human conflict

During the last decades, the human impact on ecosystems has become a massive threat to biodiversity, the diversity of species, their genetic foundations, the natural ecosystems in which they live and the functional relationships which they enter into, all over the globe (Mittermeier *et al.* 1998, Brooks *et al.* 2006). Mainly in the tropics, a variety of plant and animal species go extinct even before they are discovered and scientifically described (Primack and Corlett 2005). With every species lost, ancient evolutionary pathways and genetic resources disappear, a disaster from an ecological, ethical as well as economic point of view (Primack 1995). The main cause for this ecological crisis is the uninhibited exploitation of resources by the growing world population. This leads to extensive destruction and fragmentation of natural habitats, pollution and direct persecution of many endangered species (Yablokov and Ostroumov 1991, Harrison and Bruna 1999, Markovchick-Nicholls *et al.* 2008). Furthermore, invasions of alien species and effects of climate change may constitute important threats to biodiversity (Pimm 1998, Araujo and Rahbek 2006).

Especially at risk are regionally endemic or very specialised species and those with spacious habitat demands (Sinclair *et al.* 2006). If wildlife resource requirements are incompatible with human land use interests or if animals pose a direct physical threat to humans, a variety of conflicts may arise. Examples are numerous: herbivores may raid cropland as for instance do African elephants (*Loxodonta africana*) in rural Africa, or may be regarded as competitors to cattle such as the red kangaroo (*Macropus rufus*) in Australia (Jonzen *et al.* 2005, Sitati *et al.* 2005) Some of the most pronounced human-wildlife conflicts involve large carnivores (Graham *et al.* 2005, Gittleman *et al.* 2010): tigers (*Panthera tigris*) in Asia sometimes attack livestock and people, polar bears (*Ursus maritimus*) cross cities in Canada and cheetahs (*Acinonyx jubatus*) are killed for predating on wildlife and cattle by Namibian farmers (Marker *et al.* 2003, Dyck *et al.* 2007, Nugraha and Sugardjito 2009). Some human-wildlife conflicts can be mitigated by financial compensations for lost livestock or by giving an economic value to wildlife through tourism which is today an important source of income in many parts of the world (Krüger 2005, Treves *et al.* 2009).

Conflicts between people and wildlife in Europe

In Europe, high human population density and the resulting massive conversion of natural habitats into urban, agricultural and "cultural" landscapes poses a key problem for nature conservation (Schröder 1998). As a consequence of these conditions, conflicts between wildlife and human interests were numerous in the past and are still a problem in many areas. Throughout Europe, not only wolves (*Canis lupus*), brown bears (*Ursus arctos*) and Eurasian lynx (*Lynx lynx*) but also small carnivores such as wildcats (*Felis sylvestris*), Eurasian otters (*Lutra lutra*) or red foxes (*Vulpes vulpes*) were persecuted by people for thousands of years. Large carnivores were mainly seen as a threat to cattle and people, as in the rest of the world, and small carnivores were regarded as pest and competitors for taking game birds and mammals, fish or poultry (Linnell *et al.* 2000, Baker and Harris 2006, Klar *et al.* 2008, Sales-Luis *et al.* 2009)

In the second half of the 20th century, the attitude towards carnivorous species changed and many species which were formerly persecuted were given the status of legal protection. This resulted in the recovery and expansion of many threatened carnivores in Europe (Enserink and Vogel 2006). In Germany, a recent and quite controversial example of this process is the immigration of wolves into eastern Germany from Poland. Wolves became extinct in Germany more than 100 years ago, and traditional knowledge of how to avoid damage to livestock by wolves got lost since that time. General public acceptance for these wolf packs is increasing but illegal shooting of wolves in the affected areas still constitutes a problem (Reinhardt and Kluth 2007).

Challenges to raptor conservation in Europe

Birds of prey faced a similar negative attitude in Europe and were heavily persecuted for the same reasons as mammalian carnivores. Large raptors such as the golden eagle (*Aquila chrysaetos*) and vultures scavenging on cattle carcasses were thought to actively hunt livestock (Frey and Bijleveld van Lexmond 1994, Watson and Whitfield 2002). Many small raptor species such as northern goshawks (*Accipiter gentilis*) were intensively pursued by villagers because they were considered to prey on poultry (del Hoyo *et al.* 1994). In Germany, official premiums were paid for killing white-tailed eagles (*Haliaeetus albicilla*, Fischer 1982) until around 1910; this practise was also common for golden eagles and other raptors. In Norway, such rewards were paid for as many as 88,476 golden and white-tailed eagles during the second half of the 19th century alone (del Hoyo *et al.* 1994). As a result of this relentless pursuit, certain raptor species such as bearded vultures (*Gypaetus barbatus*),

Eurasian black vultures (*Aegypius monachus*) or golden eagles were extirpated or driven close to extinction in many European countries (Suetens and von Groendal 1971, del Hoyo *et al.* 1994, Watson 1997). Around the beginning of the 20th century, white-tailed eagles had disappeared from the whole of western and most of southern Europe and were at critical low numbers in most countries throughout the rest of the continent (Hauff 1998). At this time, only around 10-15 breeding pairs had survived in Germany (Hauff 2003).

With the beginning of the conservation movement in the first decades of the last century, public attitude towards several raptor species such as white-tailed eagles or golden eagles became more positive (Oehme 1961, Hauff 1998). Owing to the termination of payments of premiums for killed eagles, the introduction of regulations for some protection and the personal engagement of several foresters, the number of white-tailed eagles in Germany slowly began to recover (Fischer 1982). Nevertheless, shooting of eagles continued during this period. Total legal protection was enacted in 1934 and marked the starting point for effective white-tailed eagle conservation and the start of a proper recovery of its population size (Oehme 1961). Around 1950, the size of the German white-tailed eagle population was estimated to be 120 breeding pairs (Hauff 2003). Thereafter, the population stagnated at around this level for the next 30 years (Hauff 2003).

At the same time, raptor populations were observed to decline in many parts of the world. Some species such as the peregrine falcon (*Falco peregrinus*) became very rare which could not exclusively be explained by robbing of eggs or nestlings or other obvious factors (Cade 1985). The reason for these problems was found in a massive reduction of reproductive success. Many raptor species including white-tailed eagles were reported to break their eggs during breeding because eggshells were extraordinarily thin. Finally, the pesticide dichloro-diphenyltrichloroethane (DDT) was identified to be responsible for this phenomenon (Ratcliff 1967, Newton 1979, Grier 1982, Helander 1994, del Hoyo *et al.* 1994). With the prohibition of DDT and effective nest site protection starting in the 1970s, the German and other European white-tailed eagle populations soon increased substantially (Hauff 1998, Mizera 2002). This positive trend proceeded and today the German population is reported to be at a level of 570 breeding pairs (Hauff 2008), a great success for conservation.

The issue of lead poisoning

The German white-tailed eagle population still suffers from several anthropogenic threats. Again, a toxin is involved: lead poisoning was the main cause of death among white-tailed eagles found dead in Germany between 1996 and 2007; 23 % of 390 examined eagle carcasses were affected (Krone *et al.* 2009). Other anthropogenic causes of death are collisions with trains, electrocution, poisoning or accidents at wind power plants (Krone *et al.* 2003, Krone *et al.* 2009). The most important natural causes of death are territorial fights, parasitaemia and other infections (Krone *et al.* 2009).

Concerning lead, the eagles ingest the heavy metal when feeding on game carcasses or gut piles interspersed with lead bullet fragments or shot. Less important sources of lead are prey animals injured by lead ammunition and waterfowl with lead shot in their stomach which had been ingested as grit to support digestion (Kenntner *et al.* 2001, Langgemach *et al.* 2006, Helander *et al.* 2009). During necropsy, particles of lead bullets and shot were frequently demonstrated in crops and stomachs of dead eagles (Krone *et al.* 2009). White-tailed eagles mainly feed on fish and waterfowl (Oehme 1975, Struwe-Juhl 1996) but are also scavengers and carrion constitutes an important part of the diet, especially in winter when fish and waterfowl are less available (Nadjafzadeh *et al.* submitted), and most lead-poisoned eagles are collected (Krone *et al.* 2009). The gastric fluid of several raptors, including the white-tailed eagle, has a very low pH (Duke *et al.* 1975), accelerating the dissolution of ingested metallic particles and enhancing lead uptake into the bloodstream and its poisonous effects on several organ systems. Symptoms of lead intoxication include neurological disorders (e. g. blindness, loss of coordination), anaemia and diarrhoea (Fisher *et al.* 2006). The eagles may die within some days (acute intoxication) or within a few weeks by organ failure, suffocation and starvation (chronic intoxication; Pattee *et al.* 1981, Lumeij 1985).

Lead intoxication from hunter ammunition were not only reported for white-tailed eagles but also for several other raptor species such as golden eagles, bald eagles (*Haliaeetus leucocephalus*), Steller's sea eagles (*Haliaeetus pelagicus*) and bearded vultures (Wayland *et al.* 1999, Kim *et al.* 1999, Hunt *et al.* 2006, Kenntner *et al.* 2007, Hernandez and Margalida 2009). Probably the most famous example is the Californian condor (*Gymnogyps californianus*) which became almost extinct in the wild around 1980. The remaining individuals were captured and brought to breeding centres to produce offspring for an extensive reintroduction program (Green *et al.* 2008). Condors released into the wild during the past decade heavily suffered from lead poisoning, with the consequence that the population is currently not able to survive independently from human support. Almost all individuals have to be regularly caught to treat them against elevated lead concentrations with chelating agents (Cade 2007, Green *et al.* 2008). In response, the use of lead ammunition was officially banned in the state of California in 2008 (Petterson *et al.* 2009). In Condor foraging ranges in states outside California, public information campaigns were initiated to convince hunters to switch to lead-free ammunition (Cade 2007).

Purpose of this study

In Germany, the three main options to solve the lead-poisoning problem as an example of the classic conflict between wildlife welfare and human land use interests are: (1) the systematic removal of potentially lead-contaminated gut piles of shot game from the countryside; (2) a law requiring hunters to switch to lead-free ammunition; (3) a voluntary switch to the use of lead-free ammunition for hunting.

The first approach is easy to implement in principle but problematic as it just treats the symptoms but does not provide a sustainable solution to the problem itself. Furthermore, the lead-contaminated carcasses of moribund game mammals and birds which are injured by hunters but not found remain in the countryside and their lead-contaminated carcasses pose a threat to scavenging wildlife. Also, remains of kills by large mammalian carnivores were very likely an important food source for scavenging white-tailed eagles in the past before these carnivores became extinct in Germany. As human hunters hold the function of these carnivores in our landscapes today, the complete removal of all carrion and game remains such as gut piles from the environment would probably not be beneficial for white-tailed eagles and other scavengers.

Solving the lead problem by legislature is also not very promising. The assumption that hunters will simply switch to alternative ammunition after enacting a law is frequently considered naïve because there are well-known limits to the efficiency and practicality of law enforcement in such a case (Cade, 2009). Unenforceable legislation that is ignored by its target community will therefore not solve the actual issue of lead poisoning either.

A voluntary switch to using lead-free ammunition by hunters is often considered the most sustainable solution to the problem. However, it requires hunters to understand the issues involved and be persuaded that such a switch will not have disadvantages in other respects in the long term – it therefore requires appropriate research and the transfer of research results using appropriate channels to the target community, which is a time-consuming process and requires an extensive network of partners and recipients.

In April 2005, a meeting of stakeholders representing hunting organisations, the ammunition industry, conservationists, veterinarians and wildlife biologists was organised by the Leibniz Institute for Zoo and Wildlife Research (IZW) to discuss lead-poisoning in white-tailed eagles and identify open questions related to the problem. In the course of this meeting, a catalogue of such questions was assembled which all participants of the meeting assessed as important to answer in order to find a truly sustainable solution to the lead problem. As a result, the project "lead-poisoning in white-tailed eagles: causes and approaches for solutions" was initiated by the Leibniz Institute for Zoo and Wildlife Research and financed by the Federal Ministry of Education and Research (BMBF) using white-tailed eagles as a model species to represent all potentially affected scavenging species. Within this project I was in charge of answering questions on the distribution, space use and habitat use of white-tailed eagles. The results are presented in this thesis.

Information on home range size, ranging behaviour and habitat use of adult and territorial white-tailed eagles are scarce in the literature and are all based on visual observations (see introductions of the three manuscripts for brief reviews). Yet, issues related to the space use of the species were often used in arguments about the lead poisoning problem. For instance, it was commonly stated that territorial white-tailed eagles would use very large home ranges, cover long distances and even cross national borders during forays. The core of this argument was to point out that no conclusion can be drawn from the location where a lead-poisoned territorial eagle might be found because the source is unlikely to have been the local terrain. In order to provide solid scientific information on white-tailed eagle space use we intensively investigated key aspects of the spatial ecology of adult and territorial white-tailed eagles in northern Germany by means of telemetry.

Study sites

Our main study site for capturing and tracking white-tailed eagles was located in the north-western part of the Mecklenburg Lake District within and around the nature park *Nossentiner/ Schwinzer Heide* (Fig. 1, 2) in the German state of Meckenburg-Western Pomerania. The Lake District which pervades north-eastern Germany is of major importance for the survival of the German white-tailed eagle population, since it was the last refuge of the species during the period of extensive persecution (Oehme 1961). Today, the Lake District still hosts the core of the German white-tailed eagle population. The nature park *Nossentiner/Schwinzer Heide* covers an area of 365 km^2 and holds one of the highest densities of white-tailed eagles in Germany (Hauff and Mizera 2006). Currently, around 16 territorial pairs breed within its borders. Environmental conditions in this part of Germany are favourable for white-tailed eagles owing to the many water bodies which provide the main food and the comparatively low human population density. Land use is mostly extensive; the area is characterised by

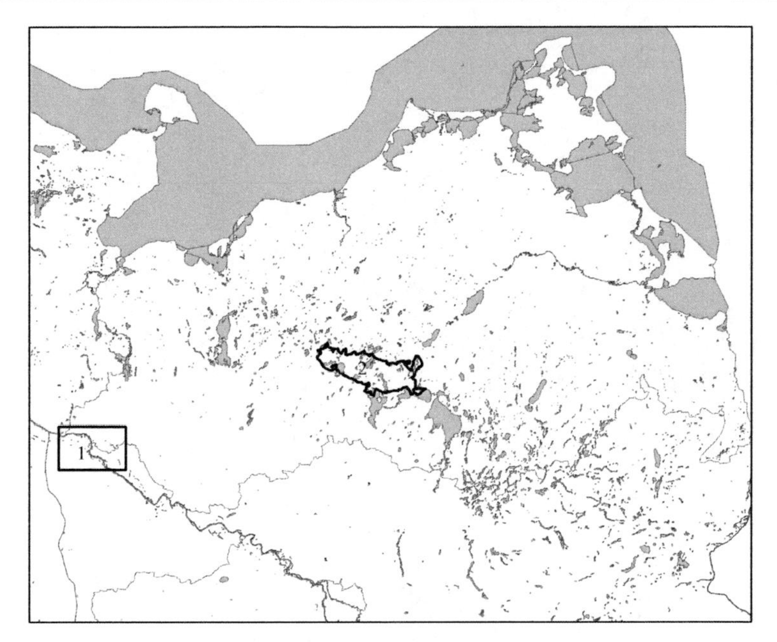

Fig. 1. The lake district of north-eastern Germany, grey: water bodies, 1: study site within the biosphere reserve *Niedersächsische Elbtalaue*, 2: nature park *Nossentiner/Schwinzer Heide*

managed forests, livestock pastures and some non-industrial agriculture (Fig. 2). Tourism is an important source of income in this region.

The other study site where only one individual was tracked comprised the northern part of the biosphere reserve *Niedersächsische Elbtalaue* (Fig. 1, Appendix). This area is less forested than the *Nossentiner/Schwinzer Heide* and dominated by livestock pastures. Alongside the river Elbe, many dikes and channels are characteristic landscape features. This section of the Elbe was re-occupied by white-tailed eagles within the last 10-15 years and is located rather close to the western margin of the current eagle population. The density of breeding pairs is comparatively low. The biosphere reserve is also a popular destination for ecotourism.

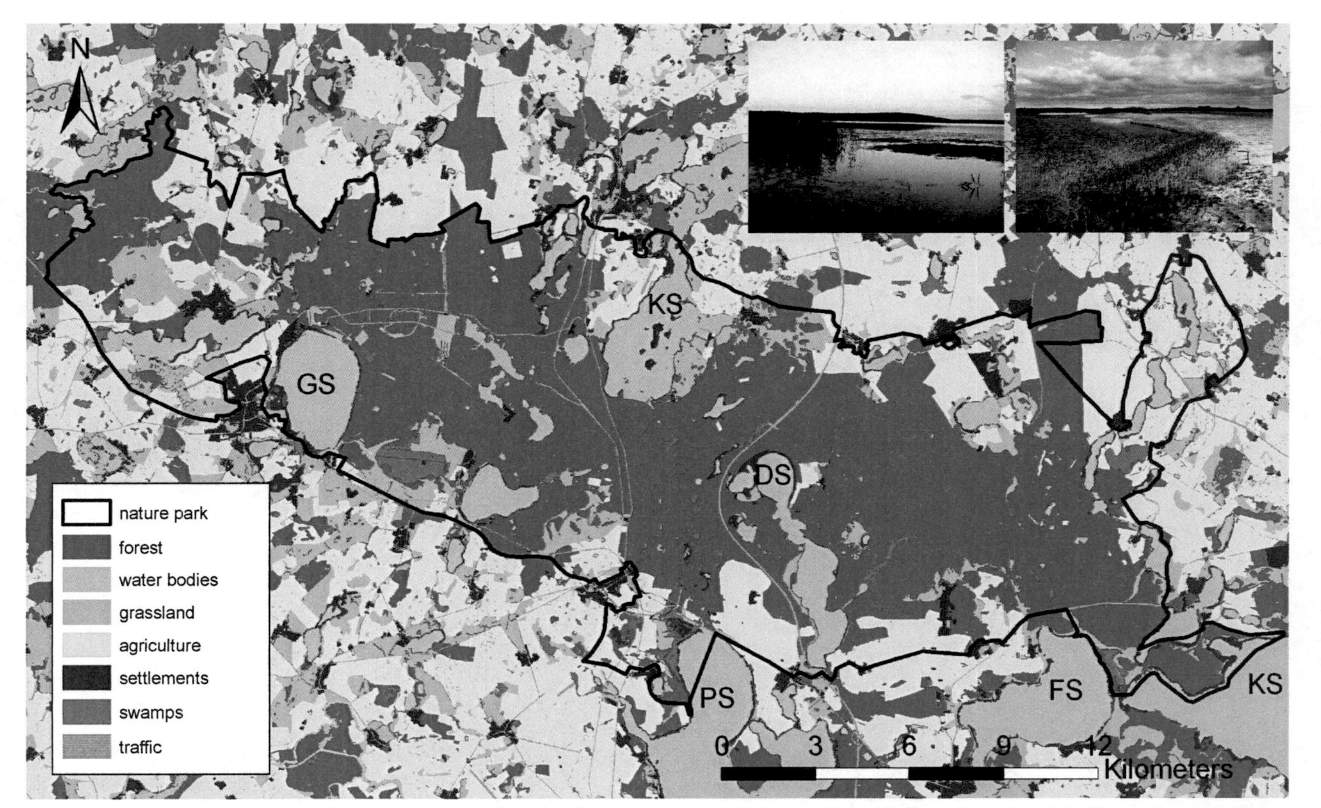

Fig. 2. The nature park *Nossentiner/Schwinzer Heide*, GS: Goldberger See, KS: Krakower See, DS: Drewitzer See, PS: Plauer See, FS: Fleesensee, KS: Kölpinsee

GPS telemetry of individual eagles

Since white-tailed eagles are highly mobile and thus difficult to track by sight, we used global positioning system (GPS) telemetry to monitor the movements of our study animals. The GPS transmitters were attached to the back of eight eagles with Teflon ribbon (Fig. 3). Transmitters weighed 170 g which accounted for approximately 3 % of body mass; a value commonly considered as acceptable for birds (Withey *et al.* 2001, Kenward 2001). No obvious effect of the transmitters on behaviour was observed throughout the study. In addition to the GPS receivers, conventional VHF elements were integrated into the transmitters. Thus it was possible to directly track eagles in the field using an antenna. All GPS data were imported into a geographic information system (GIS) for further analyses.

Fig. 3. GPS transmitter

Structure of the thesis

The results of this study are presented in three manuscripts in chapters two to four.

The focus of <u>chapter two</u> ("Roaming through the neighbourhood: ranging behaviour and extraterritorial movements of white-tailed eagles (*Haliaeetus albicilla*)") is on the detailed characterisation of space use patterns. The size and shape of home ranges are some of the most basic and important ecological attributes of a species (Sinclair *et al.* 2006). By using GPS data of the eight study animals, I examined home range sizes, seasonal changes in home range use as well as the extent and duration of extraterritorial forays. With regard to the lead poisoning problem, the investigation of the general ranging behaviour of the tracked eagles allowed a rough assessment of the possible range in which a territorial eagle can be affected by lead sources.

<u>Chapter three</u> ("Raptors in highly altered landscapes: Patterns of habitat selection and their implications for the conservation of the recovering white-tailed eagle (*Haliaeetus albicilla*) population in Germany") presents results of an extensive investigation of the habitat use of the tracked white-tailed eagles. Habitat use was compared with habitat availability on two spatial scales. Since Europe, and Germany in particular, are mainly characterised by anthropogenically altered landscapes, wildlife conservation today is often concentrated on the protection and preservation of habitats identified as important for target species. Thus, information on habitat selection patterns is a prerequisite for successful white-tailed eagle management. The consumption of potentially lead-contaminated carrion or gut piles by white-

tailed eagles is related to the use of those habitats where these food sources become available. Therefore, habitat selection by the study animals is discussed with respect to its consequences for the uptake of lead.

Finally, I developed a habitat model in order to generate a habitat suitability map for the entire land area of Germany to assess the amount of habitat available for the expanding white-tailed eagle population. This work is topic of chapter four ("Predicting suitable sites for recolonisation by a recovering and expanding raptor population: white-tailed eagles in a central European landscape"). The habitat suitability map provided in this chapter identifies sites of high probability of future occupation by dispersing white-tailed eagles. The map constitutes an important tool for conservation since it identifies areas which should be protected from major habitat alterations in order to enhance white-tailed eagle re-expansion. Furthermore, the habitat suitability map not only predicts possible future recolonisation sites but also shows where the use of lead ammunition will very likely pose a problem to white-tailed eagles within the coming decades.

In chapter five („General discussion") the results of all three manuscripts are summarised and discussed from a broader perspective. The discussion is focused on the importance of the results of my study for the conservation of white-tailed eagles and its contribution to the explanation of patterns of uptake of environmental lead.

References

Araujo MB and Rahbek C (2006). How does climate change affect biodiversity? *Science* 313: 1396-1397.

Baker PJ and Harris S (2006). Does culling reduce fox (*Vulpes vulpes*) density in commercial forests in Wales, UK? *European Journal of Wildlife Research* 52: 99-108.

Brooks TM, Mittermeier RA, da Fonseca GAB, Gerlach J, Hoffmann M, Lamoreux JF, Mittermeier CG, Pilgrim JD, and Rodrigues ASL (2006). Global biodiversity conservation priorities. *Science* 313: 58-61.

Cade TJ (2007). Exposure of California condors to lead from spent ammunition. *Journal of Wildlife Management* 71: 2125-2133.

Cade TJ (1985) Peregrine recovery in the United States. In: Newton I, Chancellor RD (eds.) *Conservation studies on raptors*, pp. 331-341. International Council for Bird Preservation, Cambridge, UK.

del Hoyo J, Elliot A, and Sargatal J (1994) Family Accipitridae (hawks and eagles). In: de Hoyo J, Elliot A, Sargatal J (eds.) *Handbook of the birds of the world: vultures to guineafowl*, pp. 52-105. Lynx Edicions, Barcelona, Spain.

Duke GE, Jegers AA, Loff G, and Evanson OA (1975) Gastric digestion in some raptors. *Comparative Biochemistry and Physiology* 50: 649-656.

Dyck MG, Soon W, Baydack RK, Legates DR, Baliunas S, Ball TF, and Hancock LO (2007) Polar bears of western Hudson Bay and climate change: Are warming spring air temperatures the "ultimate" survival control factor? *Ecological Complexity* 4: 73-84.

Enserink M and Vogel G (2006) Wildlife conservation - the carnivore comeback. *Science* 314: 746-749.

Fischer W (1982) Die Seeadler. A. Ziemsen Verlag, Wittenberg Lutherstadt, Germany.

Fisher IJ, Pain DJ, and Thomas VG (2006) A review of lead poisoning from ammunition sources in terrestrial birds. *Biological Conservation* 131: 421-432.

Frey H and Bijleveld van Lexmond M (1994). The reintroduction of the bearded vulture *Gypaetus barbatus aureus* into the Alps. In: Meyburg BU (ed.) *Raptor conservation today*, pp. 459-464. World Working Group on Birds of Prey and Owls, Pica Press, Berlin.

Gittleman JL, Funk SM, MacDonald D, and Wayne RK (2010). Carnivore conservation. Cambridge University Press, Cambridge, UK.

Graham K, Beckerman AP, and Thirgood S (2005). Human-predator-prey conflicts: ecological correlates, prey losses and patterns of management. *Biological Conservation* 122: 159-171.

Green RE, Hunt WG, Parish CN, and Newton I (2008). Effectiveness of action to reduce exposure of free-ranging California condors in Arizona and Utah to lead from spent ammunition. *Plos One* 3: e4022. doi:10.1371/journal.pone.0004022

Grier JW (1982). Ban of DDT and the subsequent recovery of reproduction in bald eagles. *Science* 218: 1232-1235.

Harrison S and Bruna E (1999). Habitat fragmentation and large-scale conservation: what do we know for sure? *Ecography* 22: 225-232.

Hauff P (1998). Bestandsentwicklung des Seeadlers *Haliaeetus albicilla* in Deutschland seit 1980 mit einem Rückblick auf die vergangenen 100 Jahre. *Vogelwelt* 119: 47-63.

Hauff P (2008). Seeadler erobert weiteres Terrain. *Nationalatlas aktuell 1 (01/2008)* URL: http://NADaktuell.ifl-leipzig.de. Leipzig, Leibniz-Institut für Länderkunde (ifL).

Hauff, P (2003). Sea eagles in Germany and their population growth in the 20th century. In: Helander B, Marquiss M, Bowerman W (eds.) *Sea Eagle 2000. Proceedings from the*

International Sea Eagles Conference in Bjorko, Sweden, 13-17 September 2000, pp. 211-218. Swedish Society for Nature Conservation/SNF, Stockholm, Sweden

Hauff P and Mizera T (2006). Verbreitung und Dichte des Seeadlers *Haliaeetus albicilla* in Deutschland und Polen: eine aktuelle Atlas Karte. *Vogelwelt* 44: 134-136.

Helander B (1994). Productivity in relation to residue levels of DDE in the eggs of white-tailed eagles *Haliaeetus albicilla* in Sweden. In: Meyburg BU, Chancellor RD (eds.) *Raptor conservation today,* pp. 735-738. World Working Group on Birds of Prey and Owls, Pica Press, Berlin.

Helander B, Axelsson J, Borg H, Holm K, and Bignert A (2009). Ingestion of lead from ammunition and lead concentrations in white-tailed sea eagles (*Haliaeetus albicilla*) in Sweden. *Science of the Total Environment* 407: 5555-5563.

Hernandez M and Margalida A (2009). Assessing the risk of lead exposure for the conservation of the endangered Pyrenean bearded vulture (*Gypaetus barbatus*) population. *Environmental Research* 109: 837-842.

Hunt WG, Burnham W, Parish CN, Burnham KK, Mutch B, and Oaks JL (2006). Bullet fragments in deer remains: Implications for lead exposure in avian scavengers. *Wildlife Society Bulletin* 34: 167-170.

Jonzen N, Pople AR, Grigg GC, and Possingham HP (2005). Of sheep and rain: large-scale population dynamics of the red kangaroo. *Journal of Animal Ecology* 74: 22-30.

Kenntner N, Crettenand Y, Fünfstück HJ, Janovsky M, and Tataruch F (2007). Lead poisoning and heavy metal exposure of golden eagles (*Aquila chrysaetos*) from the European Alps. *Journal für Ornithologie* 148: 173-177.

Kenntner N, Tataruch F, and Krone O (2001). Heavy metals in soft tissue of white-tailed eagles found dead or moribund in Germany and Austria from 1993 to 2000. *Environmental Toxicology and Chemistry* 20: 1831-1837.

Kenward RE (2001). A manual for wildlife radio tagging. Academic Press, London, UK.

Kim EY, Goto R, Iwata H, Masuda Y, Tanabe S, and Fujita S (1999). Preliminary survey of lead poisoning of Steller's sea eagle (*Haliaeetus pelagicus*) and white-tailed sea eagle (*Haliaeetus albicilla*) in Hokkaido, Japan. *Environmental Toxicology and Chemistry* 18: 448-451.

Klar N, Fernandez N, Kramer-Schadt S, Herrmann M, Trinzen M, Büttner I, and Niemitz C (2008). Habitat selection models for European wildcat conservation. *Biological Conservation* 141: 308-319.

Krone O, Kenntner N, and Tataruch F (2009). Gefährdungsursachen des Seeadlers (*Haliaeetus albicilla* L. 1758). *Denisia* 27: 139-146.

Krone O, Langgemach T, Sömmer P, and Kenntner N (2003). Causes of mortality in white-tailed sea eagles from Germany. In: Helander B, Marquiss M, Bowerman W (eds.) *Sea Eagle 2000. Proceedings from the International Sea Eagles Conference in Bjorko, Sweden, 13-17 September 2000,* pp. 211-218. Swedish Society for Nature Conservation/SNF, Stockholm, Sweden.

Krüger O (2005). The role of ecotourism in conservation: panacea or Pandora's box? *Biodiversity and Conservation* 14: 579-600

Langgemach T, Kenntner N, Krone O, Müller K, and Sömmer P (2006). Anmerkungen zur Bleivergiftung von Seeadlern (*Haliaeetus albicilla*). *Natur und Landschaft* 81: 320-326.

Linnell JDC, Swenson JE, and Andersen R (2000). Conservation of biodiversity in Scandinavian boreal forests: large carnivores as flagships, umbrellas, indicators, or keystones? *Biodiversity and Conservation* 9: 857-868.

Lumeij J (1985). Clinicopathologic aspects of lead poisoning in birds: a review. *The Veterinary Quarterly* 7: 133-138.

Marker LL, Mills MGL, and Macdonald DW (2003). Factors influencing perceptions of conflict and tolerance toward cheetahs on Namibian farmlands. *Conservation Biology* 17: 1290-1298.

Markovchick-Nicholls L, Regan HM, Deutschman DH, Widyanata A, Martin B, Noreke L, and Hunt TA (2008). Relationships between human disturbance and wildlife land use in urban habitat fragments. *Conservation Biology* 22: 99-109.

Mittermeier RA, Myers N, Thomsen JB, da Fonseca GAB, and Olivieri S (1998). Biodiversity hotspots and major tropical wilderness areas: Approaches to setting conservation priorities. *Conservation Biology* 12: 516-520.

Mizera T (2002). Bestandsentwicklung und Schutz des Seeadlers (*Haliaeetus albicilla*) in Polen im Verlauf des 20. Jahrhunderts. *Corax* 19: 85-91.

Newton I (1979). Population ecology of raptors. T. & A. D. Poyser, London, UK.

Nugraha RT and Sugardjito J (2009). Assessment and management options of human-tiger conflicts in Kerinci Seblat National Park, Sumatra, Indonesia. *Mammal Study* 34: 141-154.

Oehme G (1961). Die Bestandsentwicklung des Seeadlers, *Haliaeetus albicilla* (L.), in Deutschland mit Untersuchungen zur Wahl der Brutbiotope. In: Schildmacher H (ed.) *Beiträge zur Kenntnis deutscher Vögel,* pp. 1-61. Gustav Fischer Verlag, Jena.

Oehme G (1975). Zur Ernährungsbiologie des Seeadlers (*Haliaeetus albicilla*), unter besonderer Berücksichtigung der Populationen in den drei Nordbezirken der Deutschen Demokratischen Republik. PhD thesis, Universität Greifswald, Germany.

Pattee OH, Wiemeyer SN, Mulhern BM, Sileo L, and Carpenter JW (1981). Experimental lead-shot poisoning in bald eagles. *Journal of Wildlife Management* 45: 806-810.

Petterson JR, Sorenson K, VanTassell C, Burnett J, Scherbinski S, Welch A, and Flannagan S (2009). Blood lead concentrations in California condors released at Pinnacles National Monument, California. In: Watson RT, Fuller MR, Pokras M, Hunt G (eds.) *Ingestion of lead from spent ammunition: Implications for wildlife and humans.* pg. 238, The Peregrine Fund, Boise, Idaho, USA

Pimm SL (1998). Extinction. In: Sutherland, WJ (ed.) *Conservation science and action.* pp. 20-38, Blackwell Science, Oxford, UK.

Primack R (1995). Naturschutzbiologie. Spektrum Akademischer Verlag, Heidelberg, Germany.

Primack R and Corlett R (2005). Tropical rain forests: an ecological and biogeographical comparison. Blackwell Publishing, Oxford, UK.

Ratcliff D (1967). Decrease in eggshell weight in certain birds of prey. *Nature* 215: 208-210.

Reinhardt I and Kluth G, (2007). Leben mit Wölfen: Leitfaden für den Umgang mit einer konfliktträchtigen Tierart in Deutschland. Bundesamt für Naturschutz, Bonn, Germany.

Saito K (2009). Lead-poisoning of Steller's sea eagle (*Haliaeetus pelagicus*) and white-tailed eagle (*Haliaeetus albicilla*) caused by the ingestion of lead bullet and slugs, in Hokkaido, Japan. In: Watson RT, Fuller MR, Pokras M, Hunt G (eds.) *Ingestion of lead from spent ammunition: Implications for wildlife and humans.* pp. 302-309, The Peregrine Fund, Boise, Idaho, US

Sales-Luis T, Freitas D, and Santos-Reis M (2009). Key landscape factors for Eurasian otter *Lutra lutra* visiting rates and fish loss in estuarine fish farms. *European Journal of Wildlife Research* 55: 345-355.

Sinclair ARE, Fryxell JM, and Caughley G (2006). Wildlife ecology, conservation, and management. 2nd edition. Blackwell Science, Oxford.

Sitati NW, Walpole MJ, and Leader-Williams N (2005). Factors affecting susceptibility of farms to crop raiding by African elephants: using a predictive model to mitigate conflict. *Journal of Applied Ecology* 42: 1175-1182.

Struwe-Juhl B (1996). Brutbestand und Nahrungsökologie des Seeadlers *Haliaeetus albicilla* in Schleswig-Holstein mit Angaben zur Bestandsentwicklung in Deutschland. *Vogelwelt* 117: 341-343.

Suetens W and von Groendal P (1971). *Aegypius monachus* - Möchsgeier, Kuttengeier. In: Glutz von Blotzheim UN, Bauer K and Bezzel E (eds.), pp. 259-273. *Handbuch der Vögel Mitteleuropas, Band 4: Falconiformes*. Verlagsgesellschaft, Frankfurt/ Main, Germany.

Treves A, Jurewicz RL, Naughton-Treves L, and Wilcove DS (2009). The price of tolerance: wolf damage payments after recovery. *Biodiversity and Conservation* 18: 4003-4021.

Watson J (1997). The golden eagle. T. & A.D. Poyser, London, UK

Watson J and Whitfield P (2002). A conservation framework for the golden eagle (*Aquila chrysaetos*) in Scotland. *Journal of Raptor Research* 36: 41-49.

Wayland M, Neugebauer E, and Bollinger T (1999). Concentrations of lead in liver, kidney, and bone of bald and golden eagles. *Archives of Environmental Contamination and Toxicology* 37: 267-272.

Withey JC, Bloxton TD, and Marzluff JM (2001). Effects of tagging and location error in wildlife radiotelemetry studies. In: Millspaugh JJ, Marzluff JM. (eds.) *Radio tracking and animal populations*. pp. 45-75, Academic Press, London, UK.

Yablokov AV and Ostroumov SA (1991). Conservation of living nature and resources: problems, trends and prospects. Springer Verlag, Berlin Heidelberg, Germany.

CHAPTER 2

Roaming through the neighbourhood: ranging behaviour and extraterritorial movements of white-tailed eagles (*Haliaeetus albicilla*)

Abstract

We studied ranging behaviour and extraterritorial excursions in territorial adult white-tailed eagles (*Haliaeetus albicilla*), a raptor still in the process of recovery after severe population declines in many European countries, by means of GPS telemetry. The eight tracked adults used home ranges with an average size of 15 km^2. All study animals undertook extraterritorial movements of variable length and duration. Such movements are frequently excluded from analysis in studies of home range use. Yet in territorial species, extraterritorial movements provide important insights into the behavioural ecology of a species, as territory owners risk injuries when they embark on such excursions into neighbouring territories. Extraterritorial movements constituted an important component of the ranging behaviour of several eagles, yet they were restricted to excursions less than 39 km, despite the capacity of these raptors to range much further, and an average duration of absence of one to three days. We compared these movements with optimality models to identify the trade-offs associated with the observed ranging patterns. These identified the risk of losing territory ownership during absences as the main factor limiting excursion distance and duration. Our results showed that (1) home ranges were rather small compared to literature and (2) extraterritorial movements can also be important in territorial species such as white-tailed eagles where the main source of natural mortality are injuries from territorial fights.

Keywords: extraterritorial movements, *Haliaeetus albicilla*, home range, raptor, spatial use, telemetry, white-tailed eagle

Introduction

Information on home range size and movement patterns of a certain species are of particular importance for conservation and management (Mladenoff *et al.* 1999, Schadt *et al.* 2002, Sinclair *et al.* 2006). However, data concerning these issues are scarce for white-tailed eagles (*Haliaeetus albicilla*) and are all based on visual observations. Home range sizes were assessed by Oehme (1975) and Struwe-Juhl (1996b, 2000) for some German breeding territories and by Ganusevich (1996) for several eagles inhabiting the Kola Peninsula/Russia.

The average home ranges examined this way varied between 35 km^2 and 70 km^2. Krone *et al.* (2009a) presented preliminary results of satellite telemetry data using the Global Positioning System (GPS) for one female white-tailed eagle in northern Germany: this animal used an area of 8.2 km^2 (95 % minimum convex polygon).

Investigations of home range sizes are often topic of ecological studies but generally less attention is paid to extraterritorial movements (excursions). Extraterritorial movements occur in the context of mating (Salsbury and Armitage 1994, Norris and Stutchbury 2001, Lovari *et al.* 2008, Humbird and Neudorf 2008) or may represent pre-dispersal exploratory forays (van Ballenberghe 1983, Messier 1985). Such movements may also often be triggered by seasonally changes in resource availability, e.g. fruit masting events (Buij *et al.* 2002) or periods of food or water scarcity (Messier 1985, Hofer and East 1993, Scholz and Kappeler 2004, Frame *et al.* 2004). Extraterritorial movements are often considered to be outliers and thus not included in analyses or definitions of home range use, thereby ignoring important information about spatial ecology and resource requirements. This treatment of such data is puzzling, as in territorial species such forays outside the home range are expected to be associated with costs and thus unlikely to occur unless they are associated with appropriate benefits. For instance, costs in terms of energy and time, risk of injury from attacks by conspecific territory holders or increased predation risk are expected to influence movement behaviour (Steudel 2000, Humbird and Neudorf 2008). Furthermore, territorial species face the risk of losing their territory to intruders when moving away (Mitani and Rodman 1979, Krebs 1982).

Whilst a 'territory' is commonly defined by active or passive defence of its borders (Brown 1964, Maher and Lott 1995), there are numerous controversial concepts about what to consider to represent an animal's 'home range'. In particular, there is no consensus on how to deal with extraterritorial movements (excursions) when calculating home range areas. Whereas Burt (1943) regarded a home range as the "area traversed by the individual in its normal activities of food gathering, mating and caring for young", Jolly (1972) defined it as the area used by the animal throughout its lifetime. Hansteen (1997) pointed out the need to add a temporal dimension by defining a home range as "the area traversed by an animal during a given time period". Baker (1978) considered the biological importance of excursions and distinguished the area an animal crosses throughout its lifetime as the 'familiar area', whereas the 'home range' is considered to be the portion of the familiar area used during a given time period. Following this reasoning, the familiar area should include occasional excursions which are excluded when calculating the 'normal' home range.

In this paper we will use both concepts depending on whether excursions are of interest or not. The terms 'excursion', 'extraterritorial movement' and 'foray' are used as synonyms and describe temporary movements of an animal leaving its territory for certain time periods and distances and then returning to its original territory (Ballard *et al.* 1997, Nicholson *et al.* 2007). We studied home range use and extraterritorial movements in white-tailed eagles (*Haliaeetus albicilla*), a raptor species once almost driven to extinction by human persecution throughout central Europe. During the last 100 years, the population recovered well in Germany owing to extensive conservation efforts (Fischer 1982, Hauff 2000, Helander and Sternberg 2002) but continues to suffer from anthropogenic threats such as lead intoxication, train accidents, electrocution and poisoning (Krone *et al.* 2009b). White-tailed eagles are found in highly seasonal habitats all over their geographical range (Fischer 1982, de Hoyo *et al.* 1994). They predominantly feed on fish, waterfowl and carrion (Oehme 1975, Struwe-Juhl 1996a, Sulkava *et al.* 1997). Owing to these food requirements, white-tailed eagles are mainly confined to lakes, rivers, swamps and shorelines (Fischer 1982). Populations are resident throughout the entire year, except for the northernmost part of the Palaearctic where white-tailed eagles tend to migrate towards southern regions in winter (del Hoyo *et al.* 1994, Ueta *et al.* 1998).

We describe the home range use of eight eagles monitored with the aid of GPS telemetry. The focus of our investigations was on a detailed characterisation of extraterritorial movements. We expected to find rather flexible space use patterns for white-tailed eagles since these raptors inhabit highly seasonal habitats and are capable of swiftly covering large distances. Ranging strategies were assumed to depend on habitat quality, population density and season. Furthermore, we predicted that excursions should be triggered by gains related to mating, territory establishment/enlargement and food acquisition and limited by costs like time and energy costs of locomotion, this risk of injuries during encounters with foreign conspecifics and the risk of losing the own territory absence during absence.

Materials & methods

Study sites

This study was mainly conducted within the nature park *Nossentiner/ Schwinzer Heide* which is part of the large lake district located in the north-eastern region of Germany. The area is characterised by a high proportion of forest (mostly *Pinus*), numerous lakes, livestock pastures and low human population density. This landscape hosts an unusually high density of breeding pairs of white-tailed eagles (Hauff and Mizera 2006, Hauff *et al.* 2007) and is often regarded as 'optimal habitat' for this species. We also worked at the Elbe river in the northern part of the biosphere reserve *Niedersächsische Elbtalaue* in northern Germany (Lower Saxony). The landscape in this region is rather open and dominated by the broad river, large livestock pastures and many dikes and small channels.

Study animals and telemetry

Between 2003 and 2008 we captured eight adult white-tailed eagles using baited bow nets (Bloom 1987) or the floating-fish method (Frenzel and Anthony 1982). We used body parts and gut piles of roadkill game and dead fish as bait which were provided by hunters and fishermen in the region. Traps were continuously monitored from a hide whilst active and the captured eagles taken from the traps as quickly as possible. No eagle was injured during capturing. Seven birds (five females, two males) were equipped with satellite transmitters in the nature park *Nossentiner Heide* and one male at the Elbe river. All eight study animals were paired and territory owners maintaining nest sites. We fitted animals with backpack transmitters (170g ≈ 3% of eagle body weight, Vectronic Aerospace, Berlin) which combined both GPS and traditional radio signal (very high frequency, VHF) technology and used data transfer via the cell phone network (Global System for Mobile Communications, GSM). Transmitters were programmed to record one (six individuals), two (one eagle) or three (one eagle) GPS positions per day during daylight in a chronologically circulating schedule with a delay of one hour per day. In this manner, data from the whole course of the day could be collected.

Data preparation and home range analysis

In a first step we screened all GPS data sets for serial spatial autocorrelation using the index proposed by Swihart and Slade (1997). When a data set considerably exceeded the index threshold of 0.6, locations were removed at constant intervals from the respective data set until the index fell below or just reached the threshold. However, this procedure had to be conducted for just one data set (three positions/ day) since the location data for the other seven eagles were juedged to be spatially independent by the Swihart and Slade (1997) criterion.

To characterise home range use we applied both the traditional minimum convex polygon (MCP) method (Mohr 1947) as well as fixed kernel estimates (Worton 1989). In our study, fixed kernel home ranges were underestimated in some cases when sample sizes were very high (see also Seaman *et al.* 1999). Furthermore, 95 % fixed kernels often produced home ranges of unrealistic shape since the water bodies used for foraging by the eagles were only partially enclosed. This reflected the habit of these raptors to perch most of the day and to normally undertake rather short hunting flights before returning to the shore. Nevertheless, we still considered kernel estimates to be a useful tool for identifying areas of special importance within the home range (Worton 1989). The advantages of the MCP method are the good comparability to other studies and its robustness with respect to spatial autocorrelation (Kernohan *et al.* 2001).

The 'familiar area' (Baker 1978) of each white-tailed eagle was defined as the 100 % MCP, including all locations and excursions. The 'home range' was defined in line with the definitions of Burt (1943) and Baker (1978) by excluding excursions and using the remaining locations to calculate a 95 % MCP. Furthermore, 95 % as well as 50 % fixed kernels were estimated based on these data, with the 50 % kernels regarded as core areas. For bandwidth selection (smoothing parameter) we chose the least-squares cross-validation (LSCV) method (Gitzen *et al.* 2006, Seaman and Powell 1996). MCP and fixed kernel estimates were generated with the Animal Movement extension (Hooge and Eichenlaub 1997) within the geographic information system (GIS) software ArcView 3.3 (ESRI Inc., Redlands, USA). Nest locations during the breeding phase were disregarded to avoid biases in favour of nest site positions and single individuals. We pooled sexes since sample size was too small to test for differences and preliminary inspection of our data did not indicate a distinctive effect of gender on the ranging behaviour of our study animals.

Extraterritorial movements

Our detailed literature survey revealed that there is no standard methodology to distinguish extraterritorial movements from conventional movements within home ranges. Often, the literature did not provide any description how extraterritorial movements were identified. Due to the high variation in the frequency of excursions among our study animals, the simple extraction of a certain percentage of outliers to identify excursion locations was not an option. In order to distinguish between extraterritorial movements and home range locations, we therefore proceeded as follows: we first calculated a 95 % fixed kernel using all locations for an individual. Next, the mean distance between all successive locations falling into the 95 % kernel shape was estimated for each bird to get an idea about the magnitude of movement distances within the 'normal' home range. The mean distances were subsequently averaged over all study animals. Then, starting at the home range centre and moving outwards, every location further away from its more interior neighbour than the double mean distance between

95 % kernel locations was defined as an excursion point together with those following up. If distant excursion locations were connected to the clustering home range locations only by a single location (a 'stepping stone'), this location was regarded as being part of the excursion as well. Forays were characterised per individual by frequency, season, maximum distance covered (of all excursions), average maximum distance covered (sum of maximum distances per excursion/ number of excursions), maximum duration (of all excursions), average duration (sum of days spent on excursions/ number of excursions), and the proportion of days spent on excursions.

Statistical analyses

For a visual comparison of extraterritorial activity between the study animals, we applied time series segmentation and calculated the net displacement of all GPS data sets as described by Dettki and Ericsson (2008). Furthermore, we performed a hierarchical cluster analysis to classify all study animals according to their space use characteristics. We applied a between-groups linkage approach using the squared Euclidean distance as distance function.

All statistics were conducted in SPSS 16 (SPSS Inc., Chicago, USA). All nonparametric tests were performed in their two-tailed versions and significance levels were given as exact p-values. 'Summer' was defined to encompass the months from April to September; 'winter' months were regarded to be the six months between October and March.

Results

Sizes of 'familiar areas' and 'home ranges'

The 'familiar areas' of the eight eagles, depicted by 100 % minimum convex polygons (MCPs) including excursions, varied between 25.7 km^2 and 1081.6 km^2 (Table 1). Variation in size clearly depended on the number and extent of excursions (Table 2). Due to this enormous variation, we do not provide an average or median for this measure.

The average size of the year-round 'home ranges' of our eight study animals was 14.6 ± 3.4 km^2 if measured by 95 % MCP (Fig. 1) and 9.2 ± 4.4 km^2 if measured by 95 % fixed kernel. We recorded a minimum home range size of 6.3 km^2 (95 % MCP) or 1.6 km^2 (95 % fixed kernel). The largest home range covered an area of 35.9 km^2 (95 % MCP) or 39.3 km^2 (95 % fixed kernel, Table 1). The animals concentrated their activities within core areas of on average 1.1 ± 0.6 km^2 (50 % fixed kernel), ranging from 0.1 km^2 to 5.6 km^2 (Table 1).

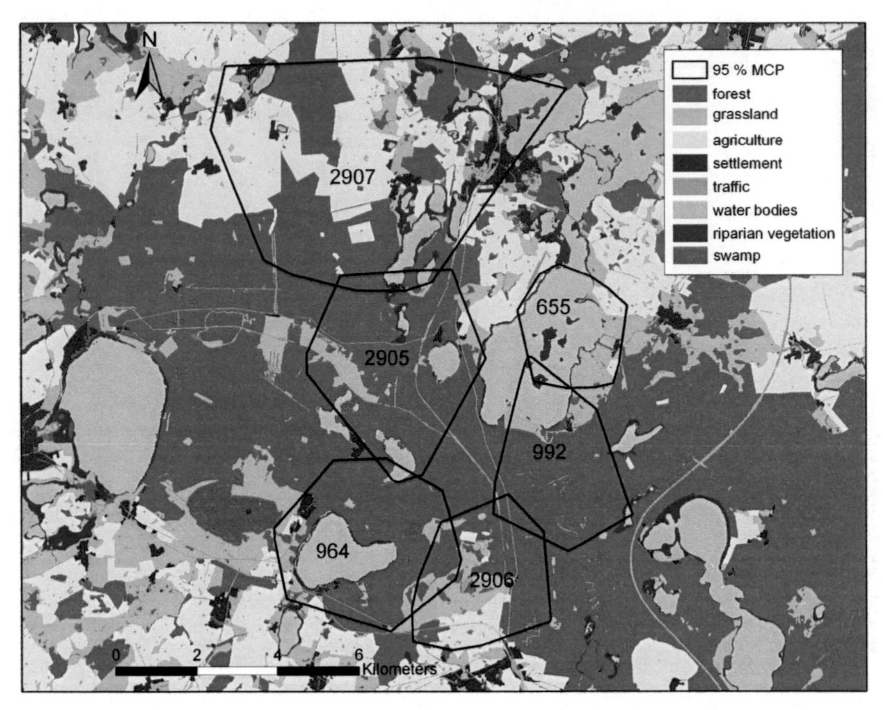

Fig. 1. Home ranges (95% MCP) of six study animals in the *Nossentiner/Schwinzer Heide*

Table 1. Sizes of all familiar areas and home ranges. MCP: Minimum Convex Polygon, N: sample size, h: bandwith value for fixed kernel analysis, underlined: eagle living at the Elbe river, * see also Krone (submitted), ** see Krone (2009a)

Animal	'familiar area' [km²]		'home range' [km²]				
	N	100% MCP	N	95% MCP	h	95% Kernel	50% Kernel
655*	571	25,7	507	6.3	173.7	1.6	0.1
472**	476	48.5	244	6.7	266.9	4.3	0.5
2907	81	62.3	81	35.9	744.8	39.3	5.6
<u>2355</u>	343	114.3	306	18.4	352.3	5.8	0.7
992*	119	310.4	105	10.2	340.6	7.5	0.9
2905	331	639.2	283	15.6	272.5	2.2	0.3
964*	263	669.7	235	14.3	263.9	6.9	0.4
2906	217	1081.6	127	9.8	288.1	5.7	0.7

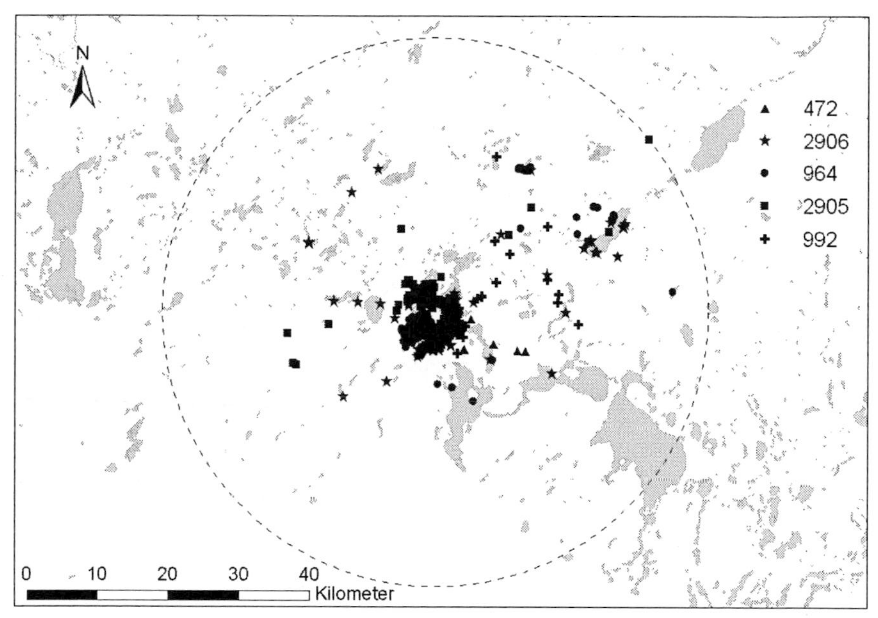

Fig. 2. Excursions of five study animals. The dotted circle represents the maximum excursion distance, grey areas depict water bodies.

The eagles used significantly larger home ranges during winter (mean 95 % MCP = 11.8 ± 1.8 km^2) than summer (mean 6.6 ± 1.7 km^2, Wilcoxon signed-rank test, N = 7, p_{exact} = 0.031).

Excursions

We observed excursions by all eight study animals. Four individuals ("655", "472", "2907", "2355") only spent between 0.4% and 2% of all observation days outside their home ranges and were not located further than approximately 12 km away from their respective home range centres (Table 2, Fig. 3a). The other four birds regularly undertook excursions of considerable length. Maximum distances were 22.8 km, 30.5 km, 36.3 km and 38.6 km as measured from the home range centres, and average excursion distances varied between 13.2 km and 23.3 km (Table 2, Fig. 2). There were pronounced differences regarding the duration of excursions and the proportion of days spent outside the home range: The longest excursion was observed for eagle "964" which once left its home range for 11 days and stayed away for 5.8 days on average (Fig. 3b). In contrast, the longest recorded absence for eagle "2906" was only five days, with on average 1.9 days spent on forays before returning home. This bird was located on extraterritorial movements on 15.2 % of 217 observation days, the highest percentage of time spent on excursions away from the 'home range' we observed (Table 2), and went more often on excursions than any other (N = 17, Fig. 3c).

Excursions were recorded in summer as well as in winter. The eagle living at the Elbe river ("2355") and eagle "992" left home ranges more often during winter than summer months, whereas the most frequently travelling eagle ("2906") showed the opposite pattern (Table 2). For the other study animals there was no distinctive seasonal pattern. Seasonal differences were either not significant or not testable because of small sample sizes.

The forays of three of the four most frequently travelling eagles ("964", "2905" and "2906") were largely directed towards lakes (72.4 %, 45.0 % and 66.6 %). The remaining extraterritorial locations were recorded in forests, forest edges, grassland as well as arable land. Three birds were located close to lakes more often in summer than during winter; movements during winter were mostly directed towards forests and agricultural land ("2905", "2906", "992").

Table 2. Excursion characteristics, $Days_{su/wi}$: observation days during summer and winter, N_{ex}: number of excursions, $N_{su/wi}$: number of excursions during summer and winter, % $Wa_{su/wi}$: percentage of excursion positions at water bodies, l_{max}: maximum excursion distance [km], l_{av}: average maximum excursion distance [km], d_{max}: maximum excursion duration [days], d_{av}: average excursion duration [days], $\%d_{ex}$: proportion of days spent on excursions in relation to the number of observation days

Animal	$Days_{su/wi}$	N_{ex}	$N_{su/wi}$	% $Wa_{su/wi}$	l_{max}	l_{av}	d_{max}	d_{av}	$\%d_{ex}$
655	111/163	1	1/0	0/-	11.7	11.7	1	1	0.4
2907	0/81	1	0/1	-/0	8.0	8.0	1	1	1.2
472	67/104	1	0/1	-/50	11.7	11.7	2	2	1.2
2355	167/179	6	2/4	-/25	10.1	6.8	2	1.2	2.0
964	106/157	5	2/3	62/92	36.3	23.3	11	5.8	11.0
2905	168/163	6	4/2	62/33	38.6	19.3	7	2.7	6.0
992	43/76	13	3/10	100/0	22.8	13.2	1	1	10.9
2906	111/106	17	10/7	90/40	30.5	21.4	5	1.9	15.2

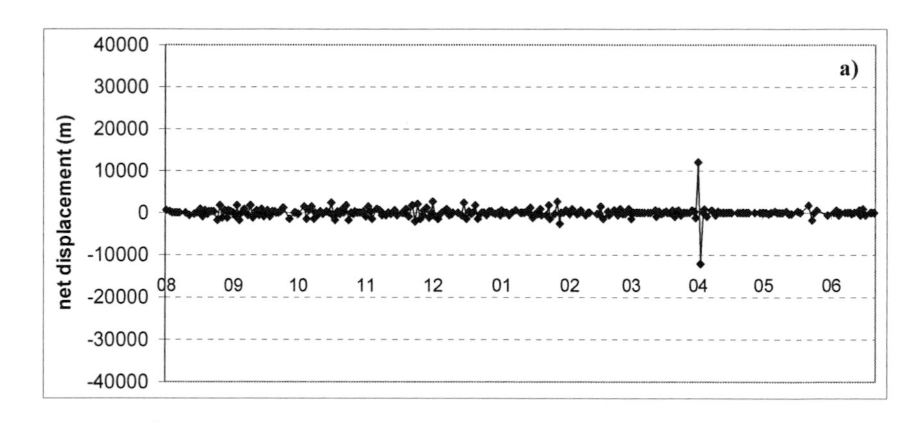

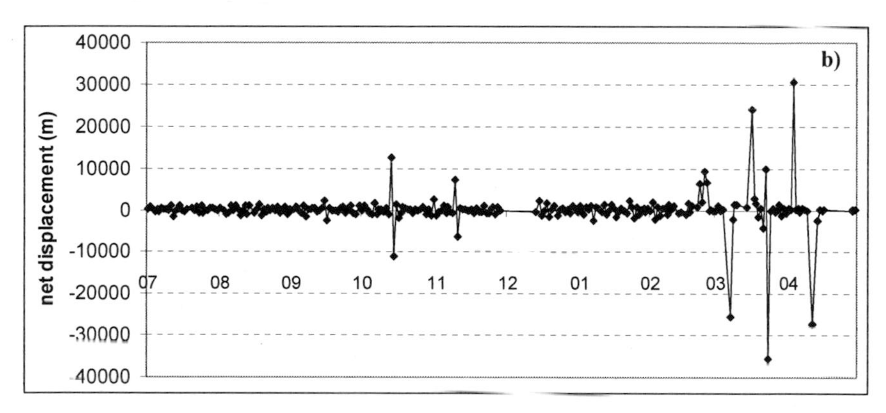

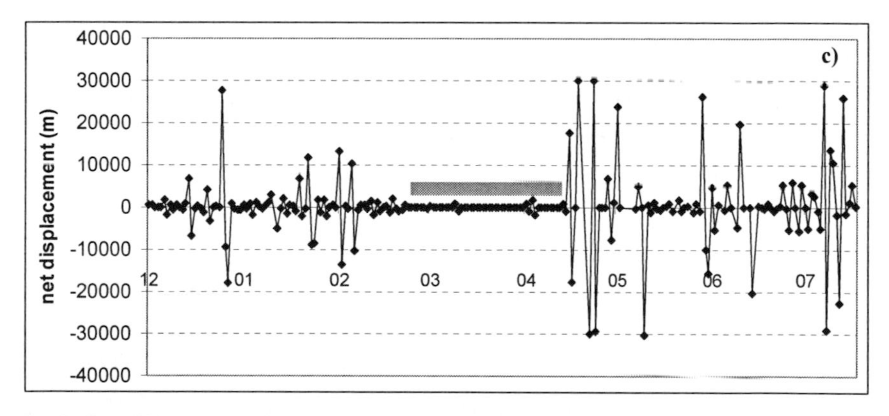

Fig. 3. The different net displacement patterns of three exemplary eagles; a) eagle "655", b) eagle "964", c) eagle "2907", grey bar indicates the time of breeding (failed attempt) for "2907", numbers underneath x axis display the month of the year

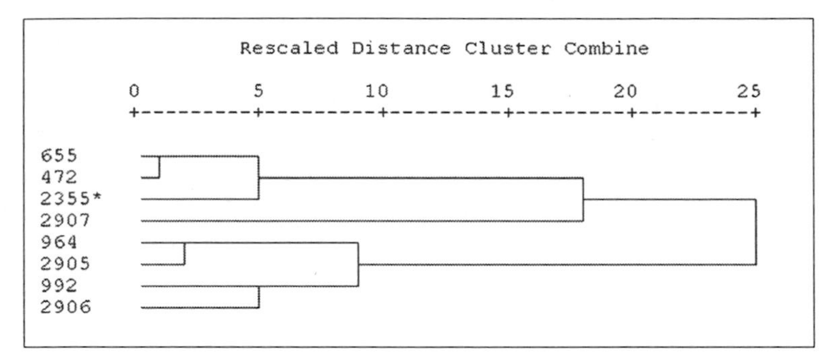

Fig. 4. Classification of all eagles by the number of observed excursions, maximum and average maximum excursion distances, average duration of forays, proportion of study days spent on excursions and size of the 95 % MCP, *: eagle living at the Elbe river

Classification

The cluster analysis used excursion parameters and home range size to group study animals as follows (Fig. 4): the four birds which undertook excursions most frequently and covered the longest distances ("964", "2905", "2906", "992") constituted one of two large clusters. Eagle "992" and "2906" as well as ""964" and "2905" formed two subgroups. Within the second main cluster, "2907" was grouped separately from the other three eagles. Furthermore, eagle "2355" stood alone, whereas "472" clustered together with "655".

Discussion

Scientific definitions and biological reality

How extraterritorial movements are defined will affect the results of home range analyses (Nicholson *et al.* 2007). The operational definition employed here clearly distinguished home range and excursion locations in all four wide-ranging eagles. For the four eagles which undertook only very rare and short forays, the impact of our definition on the classification outcome of locations was more obvious as a few isolated marginal positions defined as excursions just slightly exceeded our threshold-distance. The kind of definition chosen to identify excursions generally has to be taken into consideration when results on home range size and movement characteristics are interpreted. In our study, visual inspection of GPS data confirmed that all locations classified as extraterritorial movements were clearly separated from the main cluster of home range positions.

Sizes of 'home ranges' and 'familiar areas'

The average 'home range' size (15 km^2, 95 % MCPs) of our eight study animals was smaller than values reported by Oehme (1975), Struwe-Juhl (1996b, 2000) and Ganusevich (1996) who found average sizes of 35 km^2, 61 km^2 and 70 km^2, respectively, by visual observations. All three authors probably employed 100 % MCPs and included all observation points. Our 95 % MCP home ranges based on satellite telemetry were probably smaller than former home range estimates partly because of the effect of excluding the 5 % most marginal locations as well as extraterritorial movements. However, even the average 100 % MCP 'home range' estimate of 26 km^2 (excluding extraterritorial locations) of our study was smaller than the home range sizes published by other authors. This contrasts with the size of the 'familiar areas', using the 100 % MCPs including excursions, for which we recorded the largest total range of 1081.6 km^2 ever recorded for white-tailed eagles.

These comparisons suggest that methodical aspects such as the accuracy and efficiency of data collection as well as the choice and procedure of data analysis for the size of home ranges are mostly responsible for the reported differences in home range size, rather than ecological differences between different studies. Our sample included eagles living in a rather broad spectrum of habitats. We therefore assess our findings on the spectrum of possible home range sizes to be representative for white-tailed eagles in Germany. However, our sample included more eagles which inhabited small home ranges in rather productive and undisturbed habitats than those living in other habitats. This bias probably had an effect on the small average home range size we found.

Sizes of 'home ranges' were significantly larger during winter than during summer. This was probably an effect of the decreased fish and waterfowl availability at the water bodies in winter (Oehme 1975) which might have forced the eagles to search for food in larger areas.

Excursion characteristics of the study animals

The eagles belonged to two main groups: one comprised the four study animals undertaking only short, rare excursions; the second group consisted of the four other eagles we observed frequent and wide forays for.

The few excursions of the first group were short in distance and duration. The least roaming eagle ("655") inhabited the smallest recorded home range characterised by excellent habitat conditions. It used a large and highly productive lake in a nature reserve interspersed with several undisturbed wooded islands. The home ranges of the other three eagles were not characterised by any obvious habitat specifics. As most forays were undertaken during winter and directed towards forests and arable land, the eagles might have been in search for carrion.

Concerning the most travelling eagle ('2906'), excursions were apparently linked to food availability and habitat quality. This individual was the only study animal without direct access to a larger water body in its home range. Thus it was obviously forced to visit lakes outside its own home range within a certain distance for fishing or hunting waterfowl, thereby often intruding into the territories of other eagles.

In contrast, the home range of the second most frequently roaming eagle ("992") encompassed a large part of a productive lake with high abundances of fish and waterfowl. It undertook almost all excursions in winter and was always located on forest and agricultural land in that season. We therefore assume that this eagle left its home range in search for gut piles left by human hunters or carcasses as a response to decreased fish and waterfowl availability in winter (Nadjafzadeh *et al.* unpublished data).

For the third wide-ranging eagle ("2905") no obvious ecological factors could be identified to explain its extended forays. It was a very young adult as displayed by small black tips on some white tail feathers and iris colouration which had replaced the former female that died by lead poisoning just the year before. It is possible that the bond of this individual at its territory was not strongly developed yet and that these movements were an expression of exploratory behaviour.

The excursions of the fourth extensively travelling eagle were rather unusual; just before the transmitter ran out of battery, this male ("964") undertook three long-distance extraterritorial movements within just six weeks with the longest absences recorded for all eagles, lasting between 7 and 11 days. It started roaming around in the middle of March (breeding season) and the pair did not breed successfully in this year. It might be possible that this male was driven off by an intruding eagle in early spring and tried to get back into its home range several times. The fact that during spring significantly more eagles are killed in the course of territorial fights than during the rest of the year (Krone *et al.* 2009b) supports this suggestion.

Movement patterns in terms of optimality models

Although the purpose of extraterritorial movements were difficult to identify, there was a clear and conspicuous pattern observed in all study animals: despite the frequent extraterritorial activity of some individuals, no eagle was located further away than 39 km from the centre of its home range. Juvenile white-tailed eagles tracked during dispersal by means of GPS telemetry in the same region of Germany were easily capable of covering mean daily distances of 124 km; the maximum value recorded was 179 km day^{-1} (Marion Westphal, pers. comm.). Thus, the maximum excursion distance of 39 km was not a result of any physical constraints. Why did the eagles only move moderate distances and not further? And why was the duration of excursions limited to a few days?

In terms of optimality models the eagles have to balance a variety of gains (currencies) against several limitations (constraints) when venturing outside their home ranges (Maynard Smith 1978, Stephens and Krebs 1986). Currency might be gained food, extended spatial knowledge, monitoring 'better' territories for a chance of occupation, or meeting 'better' potential mates (van Ballenberghe 1983, Te Boekhorst 1990, Humbird and Neudorf 2008). Possible constraints include the threat of intraspecific conflicts when intruding into another territory, time and energy costs of locomotion and the risk of losing territory ownership (Mitani and Rodman 1979, Steudel 2000, Switzer *et al.* 2001).

A considerable risk of serious injuries through intraspecific conflicts is unlikely during excursions as long as the intruding eagle behaves covert and immediately retreats when the territory owner appears. The resident is always more likely to escalate fighting than an intruder because a territory owner loses the territory and perhaps also its mate in case of defeat whereas the intruder just misses a chance. Due to this asymmetry of payoff, the territory owner is more likely to win a conflict (Maynard Smith 1974, Krebs 1982). Our eagles had their own territories and probably did not intend to conquer another one whilst on excursions. Thus payoff asymmetry was substantial and the motivation of the roaming eagles to get involved into a territorial fight was certainly very low. Predation risk is commonly hypothesised to increase if an animal moves outside its home range because it has less information about hides and predators (Steudel 2000). However, as the largest European eagle, adult white-tailed eagles do not have any natural predators in Germany, so this constraint should not play a role. Most other physical threats to eagles such as lead poisoning, train collisions, electrocution or accidents at wind power plants (Krone *et al.* 2009b) are of anthropogenic origin and were not part of the evolutionary process forming the behavioural trade-off between gain and risk of extraterritorial movements.

The energetic costs of moving 20 or 30 km within a short timespan may be noticable. However, if thermic lift conditions are good and gliding over longer distances is possible, energetic costs can be considerably reduced. Constraints of time cost are strongly associated with the risk of losing territory ownership during absence. The longer a white-tailed eagle is absent, the higher the risk that an intruding bird takes over the undefended territory. Furthermore, time costs are a function of distance covered (Stephens and Krebs 1986, Alerstam and Lindström 1990).

As raptors do not scent-mark their territories like many mammal species (Johnson 1973, Hofer and East 1993,), territory ownership is mainly demonstrated to conspecifics by the omnipresence of the territory owner and territorial behaviour such as vocalisations or display flights (del Hoyo *et al.* 1994). Due to the absence of appropriate long-term marks, a territory might appear undefended and inviting for intruding eagles when the territory owner is absent. Krebs (1982) showed that in great tits (*Parus major*), territories were very quickly overtaken by intruders after the territory owners were experimentally removed. Depending on the duration of removal (0 – 24 h) and the associated payoff asymmetry, either the original territory owners or the intruders won the territorial fights which followed after the reappearance of the original owners (Krebs 1982). These issues are also likely to be important to white-tailed eagles. In areas with high eagle breeding densities and good habitat conditions, nomadic floaters in search for mates and territories are numerous. Territorial fights are the main natural cause of death among white-tailed eagles in Germany (Krone *et al.* 2009b). We observed several times that a new mate established itself in the territory of a widowed partner just within some days after an eagle died.

If movement distance and duration of extraterritorial movements are the main limiting factors influencing excursion behaviour in white-tailed eagles, then it is feasible to apply (Fig. 5) optimality models developed for optimal foraging theory. Both probably shape directly the trade-off between the gains from such movements and the risk of losing territory ownership to an intruder during absence. Range size is often regarded to be limited by economic defendability in territorial species (Brown 1964, Mitani and Rodman 1979, Peres 2000). The white-tailed eagles we studied exhibited a rather special space use pattern as some of them used both a rather small defended 'home range' (territory) and an undefended large range encompassing other eagle territories visited during excursions ('familiar area'). The maximum distance of 39 km we observed in the 50 excursions recorded in the course of our study probably represents the threshold above which excursion costs outweigh the gains. Within this range, distances are probably short enough to be quickly covered by these highly mobile birds if an eagle intends to return to its home range immediately. It is also possible that during gliding at high altitudes the own territory is still in sight within this distance. Furthermore, the eagles seemed to limit the risk of territory loss not only by excursion distance but also by

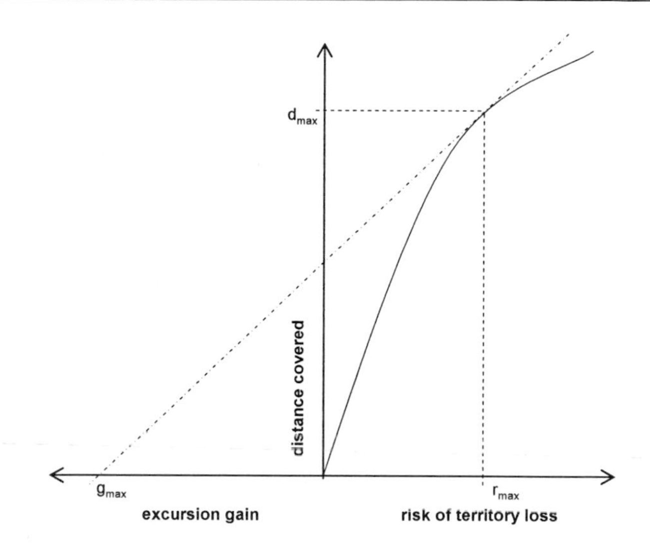

Fig. 5. The functional dependence of excursion distance on home range defendability, $d_{max} \approx 40$ km for our study animals

duration of absence. With the exception of individual "964", all other eagles left their home ranges for only one to three days on average.

In conclusion, excursive behaviour of the observed white-tailed eagle is consistent with the result predicted by an optimisation process balancing gains and risks. We suggest that the main pattern which characterises extraterritorial movements in this species is the functional dependence of excursion distance and duration on home range defendability (Fig. 5). These movements probably constitute an important part of the spatial ecology of white-tailed eagles. Consequently, not only the habitat quality of the 'home range' but also that of the surrounding landscape can be of importance for the successful establishment of a breeding territory in these raptors. Excursions are often disregarded in home range analysis. This should be reconsidered as their detailed investigation may provide important information on the behavioural and spatial ecology of a species of interest as demonstrated by this study.

References

Alerstam T and Lindström Å (1990). Optimal bird migration: the relative importance of time, energy, and safety. In: Gwinner E (ed.) *Bird migration.* pp. 331-351, Springer-Verlag, Berlin Heidelberg, Germany.

Baker RR (1978). The evolutionary ecology of animal migration. Hodder & Stoughton, London, UK.

Ballard WB, Ayres LA, Krausman PR, Reed DJ, and Fancy SG (1997). Ecology of wolves in relation to a migratory caribou herd in nothwest Alaska. *Wildlife Monographs* 135.

Bloom PH (1987). Capturing and handling raptors. In: Giron Pendleton BA, Millsap BA, Cline KW, and Bird DM (eds.). *Raptor Management Techniques Manual*. pp. 99-123, National Wildlife Federation, Washington, D.C., USA.

Brown JL (1964). The evolution of diversity in avian territorial systems. *The Wilson Bulletin* 76: 160-169.

Buij R, Wich SA, Lubis AH and Sterck EHM (2002). Seasonal movements in the Sumatran orangutan (*Pongo pygmaeus abelii*) and consequences for conservation. *Biological Conservation* 107: 83-87.

Burt WH (1943). Territoriality and home range concepts as applied to mammals. *Journal of Mammalogy* 24: 346-352.

del Hoyo J, Elliot A, and Sargatal J (1994) Family Accipitridae (hawks and eagles). In: de Hoyo J, Elliot A, Sargatal J (eds.) *Handbook of the birds of the world: vultures to guineafowl*, pp. 52-105. Lynx Edicions, Barcelona, Spain.

Dettki H and Ericsson G (2008). Screening radiolocation datasets for movement strategies with time series segmentation. *Journal of Wildlife Management* 72: 535-542.

Fischer W (1982). Die Seeadler. A. Ziemsen Verlag, Wittenberg Lutherstadt, Germany.

Frame PF, Hik DS, Cluff H, and Paquet PC (2004). Long foraging movement of a denning tundra wolf. *Arctic* 57: 196-203

Frenzel RW and Anthony RG (1982). Method for live-capturing bald eagles and osprey over open water. *United States Department of Interior, Fish and Wildlife Service, Resource Information Bulletin*: 12-13.

Ganusevich S (1996). The white-tailed sea eagle *Haliaeetus albicilla* in Kola Peninsula. *Populationsökologie Greifvogel- und Eulenarten* 4: 101-110.

Gitzen RA, Millspaugh JJ, and Kernohan BJ (2006). Bandwith selection for fixed-kernel analysis of animal utilization distibutions. *The Journal of Wildlife Management* 70: 1334-1344.

Hansteen TL, Andreassen HP, and Ims RA (1997). Effects of spatiotemporal scale on autocorrelation and home range estimators. *Journal of Wildlife Management* 61: 280-290.

Hauff P (2003). Sea eagles in Germany and their population growth in the 20th century. In: Helander B, Marquiss M, Bowerman W (eds.) *Sea Eagle 2000. Proceedings from the International Sea Eagles Conference in Bjorko, Sweden, 13-17 September 2000,* pp 211-218. Swedish Society for Nature Conservation/ SNF, Stockholm, Sweden.

Hauff P and Mizera T (2006). Verbreitung und Dichte des Seeadlers *Haliaeetus albicilla* in Deutschland und Polen: eine aktuelle Atlas Karte. *Vogelwelt* 44: 134-136.

Hauff P, Mizera T, Chavko J, Danko S, Ehmsen E, Hudec K, Probst R and Probst V (2007). Verbreitung und Dichte des Seeadlers *Haliaeetus albicilla* in sieben Ländern Mitteleuropas. *Vogelwarte* 45: 376-377.

Helander B and Sternberg T (2002). Action Plan for the conservation of white-tailed sea eagles (*Haliaeetus albicilla*). 41 pp., Birdlife International, Strasbourg, France.

Hofer H and East ML (1993). The commuting system of Serengeti spotted hyaenas: how a predator copes with migratory prey. II. Intrusion pressure and commuters' space use. *Animal Behaviour* 46: 559-574.

Hooge PN and Eichenlaub B (1997). Animal movement extension to ArcView, version 1.1. Alaska Biological Science Centre, U S Geological Survey, Anchorage, Available at http://www absc usgs gov/glba/gistools/index htm.

Humbird SK and Neudorf DL (2008). The effects of food supplementation on extraterritorial behavior in female northern cardinals. *Condor* 110: 392-395.

Johnson RP (1973). Scent-marking in mammals. *Animal Behaviour* 21: 521-535.

Jolly A (1972). The evolution of primate behaviour. Macmillan, New York, USA.

Kernohan BJ, Gitzen RA and Millspaugh JJ (2001). Analysis of animal space use and movement. In: Millspaugh JJ and Marzluff JM (eds.) *Radio tracking and animal populations*. pp. 125-166, Academic Press, London, UK

Krebs JR (1982). Territorial defence in the great tit (*Parus major*): Do residents always win? *Behavioral Ecology and Sociobiology* 11: 185-194.

Krone O, Berger A, and Schulte R (2009a). Recording movement and activity pattern of a white-tailed sea eagle (*Haliaeetus albicilla*) by a GPS datalogger. *Journal of Ornithology* 150: 273-280.

Krone O, Kenntner N, and Tataruch F (2009b). Gefährdungsursachen des Seeadlers (*Haliaeetus albicilla* L. 1758). *Denisia* 27: 139-146.

Lovari S, Bartolommei P, Meschi F and Pezzo F (2008). Going out to mate: excursion behaviour of female roe deer. *Ethology* 114: 886-896.

Maher C and Lott DF (1995). Definitions of territoriality used in the study of variation in vertebrate spacing systems. *Animal Behaviour* 49: 1581-1597.

Maynard Smith J (1974). The theory of games and the evolution of animal conflicts. *Journal of Theoretical Biology* 47: 209-221.

Maynard Smith J (1978). Optimization theory in evolution. *Annual Review of Ecology, Evolution and Systematics* 9: 31-56.

Messier F (1985). Solitary living and extraterritorial movements of wolves in relation to social status and prey abundance. *Canadian Journal of Zoology* 63: 239-245.

Mitani JC and Rodman PS (1979). Territoriality: The relation of ranging patterns and home range size to defendability, with an analysis of territoriality among primate species. *Behavioral Ecology and Sociobiology* 5: 241-251.

Mladenoff DJ, Sickley TA, and Wydeven AP (1999). Predicting gray wolf landscape recolonization: logistic regression models vs. new field data. *Ecological Applications* 9: 37-44.

Mohr CO (1947). Table of equivalent populations of North American small mammals. *American Midland Naturalist* 37: 223-249.

Nicholson KL, Ballard WB, McGee BK, and Whitlaw HA (2007). Dispersal and extraterritorial movements of swift foxes (*Vulpes velox*) in northwestern Texas. *Western North American Naturalist* 67: 102-108.

Norris D and Stutchbury BJ (2001). Extraterritorial movements of a forest songbird in a fragmented landscape. *Conservation Biology* 15: 729-736

Oehme G (1975). Zur Ernährungsbiologie des Seeadlers (*Haliaeetus albicilla*), unter besonderer Berücksichtigung der Populationen in den drei Nordbezirken der Deutschen Demokratischen Republik. PhD thesis, Universität Greifswald, Germany.

Peres CA (2000). Territorial defense and the ecology of group movements in small-bodied neotropical primates. In: Boinski S and Garber PA (eds.) *On the Move: How and Why Animals Travel in Groups*. pp. 100-125, The University of Chicago Press, Chicago, USA

Salsbury CM and Armitage KB (1994). Home range size and exploratory excursions of adult, male yellow-bellied marmots. *Journal of Mammalogy* 75: 648-656.

Schadt SA, Revilla E, Wiegand T, Knauer F, Kaczensky P, Breitenmoser U, Bufka L, Cerveny J, Koubek P, Huber T, Stanisa C and Trepl L (2002). Assessing the suitability of central European landscapes for the reintroduction of Eurasian lynx. *Journal of Applied Ecology* 39: 189-203.

Scholz F and Kappeler P (2004). Effects of seasonal water scarcity on the ranging behavior of *Eulemur fulvus rufus*. *International Journal of Primatology* 25: 599-613.

Seaman DE, Millspaugh JJ, Kernohan BJ, Brundige GC, Raedeke KJ and Gitzen RA (1999). Effects of sample size on kernel home range estimates. *Journal of Wildlife Management* 63: 739-747.

Seaman DE and Powell RA (1996). An evaluation of the accuracy of kernel density estimators for home range analysis. *Ecology* 77: 2075-2085.

Sinclair ARE, Fryxell JM and Caughley G, (2006). Wildlife ecology, conservation, and management. 2nd edition. Blackwell Science, Oxford, UK

Stephens DW and Krebs JR (1986). Foraging theory. Princeton University Press, Princeton, New Jersey, USA

Steudel K (2000). The physiology and energetics of movement: effects on individuals and groups. In: Boinski S. and Garber P. A. (eds.) *On the Move: How and Why Animals Travel in Groups*. pp. 9-23, The University of Chicago Press, Chicago, USA

Struwe-Juhl B (1996a). Brutbestand und Nahrungsökologie des Seeadlers *Haliaeetus albicilla* in Schleswig-Holstein mit Angaben zur Bestandsentwicklung in Deutschland. *Vogelwelt* 117: 341-343.

Struwe-Juhl B, (1996b). Untersuchungen zur Habitatausstattung von Seeadler-Lebensräumen in Schleswig-Holstein. Forschungsstelle Wildbiologie/ Staatliche Vogelschutzwarte Schleswig-Holstein, Kiel, Germany.

Struwe-Juhl B (2000). Funkgestützte Synchronbeobachtung - eine geeignete Methode zur Bestimmung der Aktionsräume von Großvogelarten (Ciconiidae, *Haliaeetus*) in der Brutzeit. *Populationsökologie Greifvogel- und Eulenarten* 4: 101-110.

Sulkava S, Tornberg R and Koivusaari J (1997). Diet of the white-tailed eagle *Haliaeetus albicilla* in Finland. *Ornis Fennica* 74: 65-78.

Swihart RK and Slade NA (1997). On testing for independence of animal movements. *Journal of Agricultural, Biological, and Environmental Statistics* 2: 48-63.

Switzer PV, Stamps JA and Mangel M (2001). When should a territory resident attack? *Animal Behaviour* 62: 749-759.

Te Boekhorst IJA (1990). Residential status and seasonal movements of wild orang-utans in the Gunung reserve (Sumatra, Indonesia). *Animal Behavior* 39: 1098-1109.

Ueta M, Sato F, Lobkov EG and Mita N (1998). Migration route of white-tailed eagles *Haliaeetus albicilla* in northeastern Asia. *Ibis* 140: 684-696.

van Ballenberghe V (1983). Extraterritorial movements and dispersal of wolves in southcentral Alaska. *Journal of Mammalogy* 64: 168-171

Worton BJ (1989). Kernel methods for estimating the utilization distribution in home range studies. *Ecology* 70: 164-168

CHAPTER 3

Raptors in highly altered landscapes: Patterns of habitat selection and their implications for the conservation of the recovering white-tailed eagle (*Haliaeetus albicilla*) population in Germany

Abstract

Almost everywhere throughout central Europe, wildlife has to cope with highly altered and human-dominated habitat conditions. In order to predict how wildlife populations respond to such changes, detailed knowledge on their habitat preferences would be highly desirable. We present results of the first comprehensive telemetry study on adult and territorial white-tailed eagles (*Haliaeetus albicilla*), a raptor species once almost driven to extinction in central Europe. Eight birds were fitted with GPS backpack transmitters in northern Germany in order to examine patters of habitat selection at two spatial scales in a use-availability design. Since all study animals undertook extraterritorial movements, we did not base our analyses on the area of the classic home range but developed a novel approach for defining an ecologically meaningful 'area of activity'. Selection of habitats was assessed using log-likelihood chi-square statistics on habitat selection ratios and their standard errors. Riparian vegetation was significantly preferred by the eagles at both spatial scales and together with water bodies represented the most important habitat feature. Avoided habitats were settlements, followed by arable land and traffic routes. Forest, grassland as well as swamps and small water bodies were used as expected from their availability. These findings suggest that the ongoing recovery of the European white-tailed eagle population can be directly supported by protecting riparian areas, lakes and rivers from major habitat conversions, pollution and extensive recreational activities. Our study demonstrates that the inclusion of extraterritorial movements into analyses of habitat selection is feasible and provides information suitable for conclusions on a landscape level.

Keywords: Europe, Germany, habitat selection, *Haliaeetus albicilla,* raptor conservation, white-tailed eagle

Introduction

Central Europe and especially Germany are mainly characterised by human dominated multi-use landscapes. Increasing space and resource demands for infrastructural development, recreational activity and intensive land use practices pose the main problems for modern wildlife conservation throughout the continent (Schröder 1998). Today, the protection of species with spacious habitat requirements is often focused on mitigating the effects of habitat fragmentation and alteration (Harrison and Bruna 1999, Palomares *et al.* 2000, Schadt *et al.* 2002). Therefore, knowledge on the habitat demands of a species of conservation concern is one of the prerequisites for successful management and conservation (Manly *et al.* 2002, Powell and Steidl 2002, Sinclair *et al.* 2006).

White-tailed eagles are still absent from the western part of their original European range as a result of insistent human persecution during the last centuries and were almost extirpated in Germany at the beginning of the 20th century (Hauff 2000). Today, the German eagle population is still in the process of recovery and expansion; the current population size is estimated to comprise 570 breeding pairs (Hauff 2008). Lead poisoning significantly slows population growth (Sulawa 2009) and is the most common cause of death among white-tailed eagles found dead in Germany (Kenntner *et al.* 2001, Krone *et al.* 2003). The eagles ingest lead particles by feeding on game carcasses, gut piles or prey contaminated with lead bullet remains or lead shot. Lead poisoning by spent ammunition is a widespread phenomenon, as several other scavenging raptors such as the California condor (*Gymnogyps californianus*), Steller's sea eagles (*Haliaeetus pelagicus*), bearded vultures (*Gypaetus barbatus*) or golden eagles (*Aquila chrysaetos*) (Bloom *et al.* 1989, Kim *et al.* 1999, Hunt *et al.* 2006, Cade 2007, Hernandez and Margalida 2009) suffer from it and remains a severe problem for several endangered populations (Fisher *et al.* 2006). Other anthropogenic threats problematic for white-tailed eagle management are train accidents, electrocutions or collisions with wind power plants (Krone *et al.* 2009).

White-tailed eagles were described to be associated with water bodies and undisturbed old forests in lowland as well as moderately high mountainous areas throughout Europe and northern Asia (Fischer 1982, del Hoyo *et al.* 1994). In Germany, nests are mainly built on large old trees (often pine, *Pinus sylvestris,* and beech, *Fagus sylvatica*) within moderate distances from lakes or rivers (Fischer 1982, Hauff 2001). When eagles choose the location of their nest site they avoid the proximity of houses and roads (Struwe-Juhl 1996a, Folkestad 2003). Detailed telemetry studies concerning habitat use and selection of adult and territorial white-tailed eagles are not available. We studied habitat selection by white-tailed eagles by tracking eight adult territorial individuals in northern Germany by means of Global Positioning System (GPS) telemetry. Investigations were performed within the framework of a use-availability design (Aebischer *et al.* 1993, Alldredge *et al.* 1998, Manly *et al.* 2002) at two spatial scales corresponding to Johnson's (1980) second and third order of habitat selection.

A driving force in the evolution of habitat preferences and patterns of resource selection is the increase in reproductive fitness of individuals selecting habitats in which food intake or nesting success can be maximised (Gaines *et al.* 2005). Predator avoidance is a factor which has an important impact on the selection of optimal habitats in many species (Lima 1998, Mao *et al.* 2005, Morosinotto *et al.* 2010). Most raptors are not affected by predation to a substantial extent. Accordingly, raptor densities were described to be mainly limited by physical nesting habitat or food, whichever is shorter in supply (Newton 1979). Therefore, habitat selection patterns of white-tailed eagles were hypothesised to be mainly a result of special nest site requirements, territory defence constraints and optimal food acquisition.

Foraging is always related to the usage of certain habitats. Thus, habitat selection of white-tailed eagles should reflect their common feeding habits. For instance, water bodies are expected to be habitats of major importance, as eagles mainly prey on fish and waterfowl (Oehme 1975, Struwe-Juhl 1996b). If easily available or during times of decreased fish or waterfowl accessibility, they also use carrion and sometimes hunt small mammals (Sulkava 1997, Nadjafzadeh *et al.* submitted). Therefore, eagles should also choose habitats were these alternative food resources are potentially available which are forests, grasslands or arable land. Another factor probably influencing habitat selection in white-tailed eagles is their pronounced fear of people. Habitats with a high level of expected human disturbances, such as villages or the vicinity of roads, should thus be avoided. The aim of this study was to identify favoured or avoided habitats and their ecological importance for white-tailed eagles with respect to management and conservation purposes.

Material and methods

Study area and animals

We tracked seven white-tailed eagles (two males, five females) in the nature park *Nossentiner/Schwinzer Heide* located in north-eastern Germany. The nature park is part of the Mecklenburg Lake District which hosts the core of the German eagle population (Hauff 2006). It is characterised by many water bodies, large forests, livestock pastures land and a relatively low human population density compared to the rest of Germany. A third male was monitored in the biosphere reserve *Niedersächsische Elbtalaue* along the northern part of the Elbe river. Habitat conditions in this area were very different from the nature park as the region is only sparsely forested and mainly dominated by pasture farming. During the summer months both study sites are popular destinations for green tourism.

Telemetry and GIS work

The eight study animals were fitted with backpack transmitters which combined Global Positioning System (GPS) and very high frequency (VHF) technology. They were attached to the eagles with Teflon ribbons and weighed 170 g (≤ 3 % body weight, Vectronic Aerospace, Berlin). Depending on transmitter type, GPS location data were either stored for subsequent download using ultra high frequency (UHF) or sent to our institute once per week via the Global System for Mobile Communications (GSM). The GPS elements received one (six eagles), two (one eagle) or three (one eagle) positions per day in a chronologically circulating schedule with a daily delay of one hour. The home ranges of the eight eagles covered on average 15 km^2 (95 % minimum convex polygons excluding extraterritorial movements, see chapter 2).

All GPS location data were imported into a geographical information system (ArcGIS V. 8.3, ESRI Inc., Redlands, USA) and projected on vector maps of the detailed biotope and land use surveys of the state offices for the environment of Mecklenburg Western-Pomerania and Brandenburg. Habitat selection analysis was bases on the eight habitat categories 'forest', 'grassland', 'arable land', 'settlements', 'traffic' (roads and railways), 'open water' (water bodies > 100m diameter), 'riparian vegetation' and 'swamps/small water bodies' (water bodies < 100m diameter).

Structure of GPS location data

Except for one eagle, none of the data sets showed evidence for serial spatial autocorrelation, as judged by the Swihart & Slade index (Swihart and Slade 1997). Locations were removed from the correlated data set (three positions day^{-1}) in a constant interval until the Swihart & Slade index for autocorrelation indicated spatial independence (Swihart and Slade 1997).

As we were particularly interested in habitat selection during foraging activities and not in nest site selection, nest positions of breeding study animals were counted as a single location during breeding or failed breeding attempts (between February/ March and June), in order to avoid its overrepresentation in habitat selection analyses.

All tracked birds undertook extraterritorial movements of variable length and duration (see chapter 2). For four individuals, the percentage of locations recorded on such movements was as high as maximum 15% of all locations, whilst for the other four only rare and short excursions were documented (chapter 2). In the frequently applied conventional approach to habitat selection analysis extraterritorial locations are ignored. We considered this to be inappropriate, as such locations very likely indicated the use of important habitats or resources not available within the respective eagle territories. We will explain below how we incorporated extraterritorial movements into our habitat use–habitat availability design.

Spatial scales and definitions of used and available habitat

To reflect the process of movement decision-making by the eagles as realistically as possible, habitat selection was analysed in a hierarchical design on two levels: (1) the selection of an activity range from a subset of available habitats within a wider surrounding (second order selection (Johnson 1980), and (2) the selection of habitats (GPS locations) from the habitats available within the activity range (third order selection (Johnson 1980). Habitat selection at both spatial scales was examined using the design III approach which defines the use and availability of habitats individually for every study animal (Thomas and Taylor 1990).

Since locations obtained during extraterritorial movements had to be included in the analysis, it was not sufficient to use the classic home range area as the spatial reference unit to study second as well as third order selection. We therefore proceeded as follows: For second order selection, habitat availability was defined as the habitat composition within a circle with a radius of 39 km around the centre of every home range (Fig. 1a). The radius of 39 km was based on the maximum excursion length of 39 km observed for all eagles (chapter 2). This area is likely to reflect the typical ranging capacity of eagles much better than the standard definitions of available habitat at second order scale such as study site area or the polygon area comprising all animal locations. To quantify habitat use, we created a buffer zone around each GPS location with a radius that was derived as follows: We measured the average successive distances between all GPS locations that were included in the 95 % fixed kernel home range of every eagle (Worton 1989). Fixed kernels were calculated using least-squares cross-validation (LSCV) for bandwidth selection (Gitzen *et al.* 2006). This procedure provided an estimate of the magnitude of the ‘typical’ movement distances of each eagle. Using this distance as radius to create a buffer zone around every GPS location defined a local ‘activity range’ (Fig. 1a), and the area within such ranges was categorised as used habitat for analyses of second order selection. The outcome were dense clusters of buffer

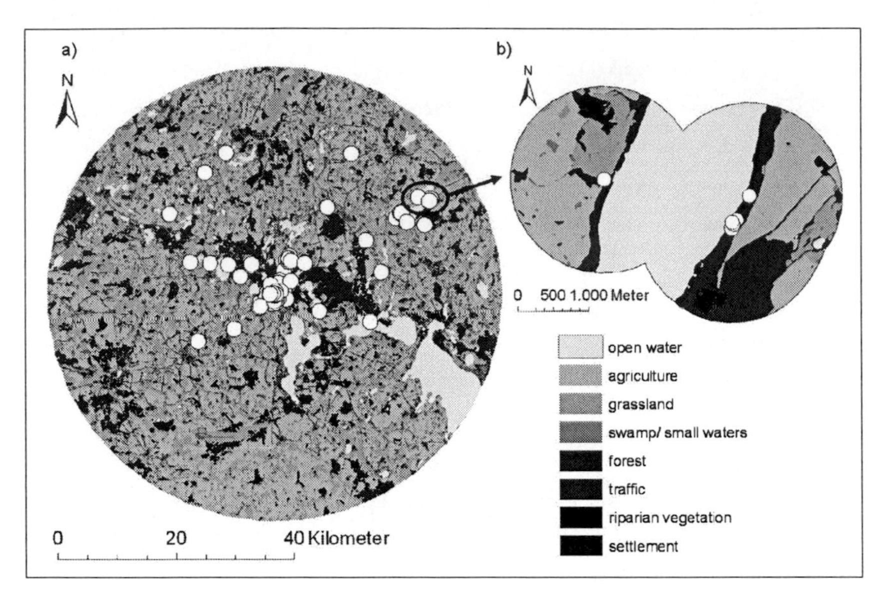

Fig. 1 Study design illustrated by the example of one study animal (eagle "2906", note the high extraterritorial activity of this individual, N = 279 GPS locations); a) study design applied at second order scale: used = white buffer zones around GPS locations (r = 1318 m) called 'activity areas', available = area within the a circle of 39 km radius around the centre of the home range; b) study design applied at third order scale: used = white buffer zones around GPS locations (r = 100 m), available = merged area of individual 'activity areas' (r = 1318 m)

zones inside the home range of every eagle and single isolated buffer zones at locations on extraterritorial movements. Buffer zones were not merged into a single shape. Thus, the frequent usage of habitat types inside the home range essential to the eagle was given a heavier weight in the analysis than the rare (but important) use of habitats visited during excursions.

For the investigation of third order selection, the merged area of individual 'activity areas' defined as used habitat for second order selection was now classified as available habitat (Fig. 1b). Used habitat was defined as the area within a small buffer zone with a radius of 100 m around every GPS location. Generating such buffer zones instead of using exact GPS positions seemed appropriate because white-tailed eagles showed a strong preference for perching at habitat edges. Since the accuracy of our GPS locations allowed an error of 15 m, frequent misclassifications of GPS locations in terms of used habitats were likely. The radius of 100 m was chosen because more than half of all eagle records were found within a distance of 100 m from larger water bodies. As our own observations demonstrated, eagles frequently used structures in the proximity of water bodies to perch and start hunting flights for fish or waterfowl; therefore this behaviour might hold true for other habitat types as well.

Statistical analyses

Numerous methods have been published to conduct statistical analysis of habitat use and preferences, such as generalized linear and additive models for estimating resource selection probability functions, compositional analysis, χ^2 goodness-of-fit tests, simple regressions, discrete-choice models, ecological niche factor analysis or distance-based approaches (Neu *et al.* 1974, Aebischer *et al.* 1993, Manly *et al.* 2002, Hirzel *et al.* 2002, Boyce *et al.* 2002, Conner *et al.* 2003, Alldredge and Griswold 2006, Thomas and Taylor 2006, Lele 2009). As the results of a study may depend on the chosen method, the choice of a particular model has to be carefully considered (McClean *et al.* 1998). Our study used the classic use-availability design and required a method that best reflected Manly et al.'s (2002) definition of selectivity: "when resources are used disproportionately to their availability, use is said to be selective". The most appropriate method was therefore a conventional goodness-of-fit test on habitat selection ratios and their standard errors for the population of study animals following the procedures suggested by Manly et al. (2002). Hence, the population selection ratio $\hat{w}_i$ was calculated for each habitat type i which took the form

$$\hat{w}_i = u_{i+} / \sum \pi_{ij} u_{+j}$$

where u_{i+} was the total number of used units in category i, π_{ij} represented the proportion of habitat available for eagle j in habitat category i and u_{+j} represented the number of habitat units used by eagle j. Summation was over all study animals. The population ratios $\hat{w}_i$ of every habitat category i were then tested for significant deviation from 1 by the comparison of

$$\{(\hat{w}_i - 1) / se(\hat{w}_i)\}^2$$

with the critical values of the χ^2 distribution with one degree of freedom (Manly *et al.* 2002). A a $\hat{w}_i$ of 1 would indicate habitat use in proportion to habitat availability, a $\hat{w}_i$ greater 1 would indicate positive habitat selection (preference) and a $\hat{w}_i$ less than 1 would indicate negative habitat selection (avoidance). We used Bonferroni-corrected significance levels to account for an increased risk of type 1 errors during multiple testing (Manly *et al.* 2002). The null hypothesis for all tests was that there was no habitat selection in terms of preferences or avoidance by the eagles.

Selection for habitat edges was investigated by generating the same number of random points within the 'activity areas' as were collected by GPS telemetry for every study animal. The distances between GPS locations and the closest habitat edges were compared with the distances between random points and the closest edges using a Wilcoxon signed-ranks test calculated in SPSS 16 (SPSS Inc., Chicago, Illinois). Seasonal differences in habitat use were also tested with a Wilcoxon signed-ranks test. Nonparametric statistics were performed two-tailed and exact P values were provided. Summer was defined as the period of April to September and winter as the six months between October and March. Data are presented as means ($\bar{x}$) ± standard error of the mean (sem).

Results

We collected between 81 and 571 GPS locations per study animal over an average period of nine months. Average distances between successive locations within the respective 95% fixed kernels varied between 726 m and 3789 m. These were the distances used as radius for the definition of buffering zones around GPS locations in order to define the 'activity ranges' of second order habitat selection analyses.

The eagles were found significantly closer to habitat edges ($\bar{x}$ = 27.9 ± 4.0 m) than expected from a random distribution ($\bar{x}$ = 69.0 ± 4.8 m, Wilcoxon signed-ranks test, N = 8, p_{exact} = 0.008). The median percentage of GPS locations found within 50 m distance from water bodies was 40.9 %. In total, 56.3 % of all records were taken not further than 100 m away from shorelines.

The eagles were located significantly closer to open water in summer ($\bar{x}$ = 341.2 ± 169.0 m, April-September) than during winter ($\bar{x}$ = 546.1 ± 216.0 m, October-March, Wilcoxon signed-ranks test, N = 7, p_{exact} = 0.016). For the other habitat types directly related to foraging such as 'forest', 'grassland' or 'agriculture', no significant seasonal differences regarding distances to GPS locations were found.

Selection at the 'activity range scale' (second order selection)

The habitat type 'open water' had the highest positive selection ratio of all habitat categories ($\hat{w}_i$ = 5.87 ± 2.06), although it was not significantly different from 1 (Table 1). The minimum proportion of 'open water' within an 'activity range' was 0.077, the maximum proportion was 0.746 (Table 1). The selection ratio of 'riparian vegetation' was similarly high

Table 1. Habitat selection parameters at the 'activity range' scale (second order selection), o_i: proportion of habitat used, π_i: proportion of habitat available, $\hat{w}_i$:selection ratio, SE: standard error, B_i: standardised selection ratio, P: p-value of χ^2 statistics, D_{sig}: direction of significant selection (Bonferroni-adjusted $\alpha = 0.0063$), -: underselected (avoided), +: overselected (preferred)

Habitat	o_i	SE (o_i)	o_{ij}max	o_{ij}min	π_i	$\hat{w}_i$	SE ($\hat{w}_i$)	B_i	χ^2	P	D_{sig}
Forest	**0.358**	0.011	0.741	0.027	**0.244**	**1.47**	0.51	0.094	0.85	0.357	
Grassland	**0.149**	0.008	0.452	0.049	**0.173**	**0.84**	0.28	0.054	0.32	0.571	
Arable land	**0.086**	0.006	0.291	0.007	**0.449**	**0.19**	0.06	0.012	165.34	<0.001	-
Settlement	**0.012**	0.002	0.031	0.002	**0.045**	**0.26**	0.07	0.017	117.10	<0.001	-
Traffic	**0.007**	0.002	0.014	0.001	**0.007**	**0.90**	0.25	0.057	0.17	0.684	
Open water	**0.324**	0.010	0.077	0.746	**0.056**	**5.87**	2.06	0.375	5.60	0.018	
Riparian vegetation	**0.054**	0.005	0.101	0.012	**0.010**	**5.34**	1.56	0.343	7.72	0.005	+
Swamp/small water bodies	**0.012**	0.002	0.031	0.005	**0.016**	**0.75**	0.30	0.048	0.70	0.402	

Table 2. Habitat selection parameters at the 'within-activity range' scale (third order selection), o_i: proportion of habitat used, π_i: proportion of habitat available, w_i:selection ratio, SE: standard error, B_i: standardised selection ratio, P: p-value of χ^2 statistics, D_{sig}: direction of significant selection (Bonferroni-adjusted $\alpha = 0.0063$), -: underselected (avoided), +: overselected (preferred)

Habitat	o_i	SE (o_i)	o_{ij}max	o_{ij}min	π_i	$\hat{w}_i$	SE (w_i)	B_i	χ^2	P	D_{sig}
Forest	**0.375**	0.011	0.659	0.027	**0.374**	**1.06**	0.26	0.095	0.05	0.829	
Grassland	**0.136**	0.008	0.340	0.057	**0.176**	**0.71**	0.11	0.064	6.43	0.011	
Arable land	**0.051**	0.005	0.173	0.008	**0.243**	**0.23**	0.06	0.021	173.81	<0.001	-
Settlement	**0.002**	0.001	0.022	0.000	**0.022**	**0.09**	0.05	0.008	322.35	<0.001	-
Traffic	**0.003**	0.001	0.012	0.000	**0.008**	**0.37**	0.12	0.033	27.05	<0.001	-
Open water	**0.241**	0.010	0.466	0.084	**0.135**	**1.53**	0.19	0.138	7.50	0.006	+
Riparian vegetation	**0.177**	0.009	0.399	0.046	**0.026**	**6.13**	1.73	0.553	8.79	0.003	+
Swamp/small water bodies	**0.016**	0.003	0.063	0.001	**0.016**	**0.98**	0.31	0.088	0.01	0.937	

and significant ($\hat{w}_i = 5.34 \pm 1.56$, N = 8, df = 1, $\chi^2 = 7.72$, p = 0.005). The habitats 'forest' ($\hat{w}_i = 1.47 \pm 0.51$), 'traffic' ($\hat{w}_i = 0.90 \pm 0.25$), 'grassland' ($\hat{w}_i = 0.84 \pm 0.28$) and 'swamp/small water bodies' ($\hat{w}_i = 0.75 \pm 0.30$) were used in proportion to their availability (Table 1). 'Settlements' within individual buffer zones were significantly avoided ($\hat{w}_i = 0.26 \pm 0.07$, N = 8, df = 1, $\chi^2 = 117.10$, p < 0.001) as was 'arable land' ($\hat{w}_i = 0.19 \pm 0.06$, N = 8, df = 1, $\chi^2 = 165.34$, p < 0.001).

Selection at the 'within-activity range' scale (third order selection)

When comparing the habitat composition of the small 100 m buffer zones around GPS locations to the composition of the 'activity ranges', the habitat 'riparian vegetation' was most strongly preferred (Table 2, $\hat{w}_i = 6.13 \pm 1.73$, N = 8, df = 1, $\chi^2 = 8.79$, p = 0.003). 'Open water' was also used significantly more than expected from availability ($\hat{w}_i = 1.53 \pm 0.19$, N = 8, df = 1, $\chi^2 = 7.50$, p = 0.006). The selection ratios for 'forest' ($\hat{w}_i = 1.06 \pm 0.26$), 'grassland' ($\hat{w}_i = 0.71 \pm 0.11$) and 'swamp/small water bodies' ($\hat{w}_i = 0.98 \pm 0.31$) showed no significant deviation from 1 (Table 2); these habitats were used in proportion to their respective availabilities. The habitat 'traffic' was significantly avoided ($\hat{w}_i = 0.37 \pm 0.12$, N = 8, df = 1, $\chi^2 = 27.05$, p < 0.001), as were the habitat types 'arable land' ($\hat{w}_i = 0.23 \pm 0.06$) and 'settlement' ($\hat{w}_i = 0.09 \pm 0.05$) (Table 2), with the latter having the lowest selection ratio of all habitats in the analysis (N = 8, df = 1, χ^2 for 'agriculture' = 173.81, χ^2 for 'settlement' = 322.35, p < 0.001 in each case).

Discussion

The eagles showed a strong preference for habitat edges. The reason for this behaviour was most likely the improved view over the terrain when perching on elevated marginal structures close to areas interesting for food acquisition or territory defence.

Selection patterns were scale-independent for some habitat categories: at both spatial scales 'riparian vegetation' was strongly favoured whereas the eagles strongly avoided 'arable land' and 'settlements'. For the other habitat types, the significance of selection was scale-dependent. For instance, the selection ratio for 'open water' was very high (but not significant) for second order selection and high and significant at the third order level. The fact that more than half of all GPS locations were taken within a distance of less than 100 m from a water body demonstrates the importance of this habitat type for white-tailed eagles. 'Forest', 'grassland' and 'swamps/small water bodies' were of less relevance and used in proportion to availability at both scales. Interestingly, 'traffic' was only avoided at the 'within-activity range' level (third order selection) but not during the establishment of the 'activity ranges' (second order selection). 'Settlements' were strongly avoided at both spatial levels. Such consistency of results across the spatial scales emphasises the strength and distinctiveness of the observed habitat selection patterns.

Patterns of habitat selection

The different habitats in the analysis are clearly either related to food acquisition or the amount of human disturbances. The availability and accessibility of food resources is a main factor driving habitat selection in many species (Oppel *et al.* 2004, Sinclair *et al.* 2006). With respect to food acquisition, the habitat category 'open water' would have the highest rank since it provides the eagles' main food. The amount of water surface within the 'activity ranges' varied tremendously among study animals, thereby producing a large standard error of the selection ratio estimate, rendering it to be non-significant in the analysis of second order selection. The lowest percentage cover of water within an activity range was 7.7 %. It is very unlikely that an activity range without any water body could support a white-tailed eagle in the long term. Closely related to the usage of water bodies for hunting fish and waterfowl was the strong and significant preference for 'riparian vegetation' across spatial scales. The eagles intensively used this habitat type for perching and observing the water surface in search for prey. GPS positions could only be recorded directly above the water surface during hunting flights or gliding which are comparatively rare activities and unlikely to be captured with our GPS location recording routine. This explains why the absolute value of the selection ratio $\hat{w}_i$ for 'open water' was much smaller but significant (owing to the small standard error) at third order selection than at the second order level, whereas 'riparian vegetation' was equally strongly selected at both scales. In conclusion, 'open water' and 'riparian vegetation' were the habitats of highest value to the eagles which is in accordance with an expected habitat ranking based on feeding habits.

The habitat categories 'forest', 'grassland' and 'swamp/small water bodies' were also ranked as potentially interesting for the eagles in terms of food because these habitats might provide carcasses, gut piles, small mammals or waterfowl. Nevertheless, they were neither preferred nor avoided but used in accordance with their availability, perhaps a reflection of the decreased importance of carrion and small mammals as food sources compared to the importance of fish and waterfowl. Another reason for the lack of preference of these habitats could be the fact that the availability of carcasses, gut piles and small mammals is generally distributed over three habitat types ('forest', 'grassland', 'arable land') which diminishes the significance of each single habitat. 'Swamps/small water bodies' are probably not visited by waterfowl in sufficient quantities interesting to white-tailed eagles. These habitats often also provide good shelter in terms of dense vegetation which hampers access to prey for the eagles.

The sixth habitat category related to food acquisition was 'arable land'. Similar to other wildlife (Marzluff *et al.* 1997, Dickson and Beier 2002, Menzel *et al.* 2005, Benson and Chamberlain 2007), our study animals distinctively avoided 'arable land' at both spatial scales. The main cause for this avoidance was probably the lack of suitable food or prey and an increased level of disturbance by people. With regard to a habitat ranking based on food acquisition, 'arable land' had a lower value to the eagles than expected. Human hunting activity might in general be lower on arable land compared to forests and grassland and thus carcasses and gut piles be less available. Furthermore, gut piles of shot game are probably rarely left in crop fields by hunters during the vegetation period and moribund game avoids open agricultural areas without any shelter.

The two habitats which explicitly represented anthropogenic sources of disturbance were 'traffic' and 'settlements'. Interestingly, 'traffic' seemed to be tolerated in the 'activity ranges' but not in the closer surrounding of perches (100 m buffer zones around GPS locations). Traffic routes often provide food in form of road kills, but frequent disturbances by vehicles might keep the eagles from using the close vicinity of roads. The second most important cause of death among white-tailed eagles in Germany are train accidents (Krone *et al.* 2003, Krone *et al.* 2009). Due to the low frequency of passing trains the eagles feed on carcasses of overrun animals close to the tracks and are then hit and killed themselves. Nevertheless, this food source is highly irregular and it is probably not very profitable for eagles to perch at railway tracks. 'Settlement' was the most strongly avoided land cover type in our study. That reflects the low rank of this land cover regarding food acquisition, since the few interesting food sources within cities or villages are not accessible to white-tailed eagles owing to their pronounced fear of people. White-tailed eagles were once almost persecuted to extinction and eagles still react very sensitive to human presence (Helander and Sternberg 2002).

In conclusion, the patterns of habitat selection observed in our study animals were consistent with the predictions based on food preferences, foraging habits and the avoidance of disturbance by people. It is difficult to disentangle the effects of food preferences from the alternative hypothesis of habitat selection for the purpose of territory defence. It is quite possible and feasible that perches used for observing surrounding areas for potential prey might in parallel and simultaneously also be important in terms of territory defence and used to monitor the eagle territory for intrusions by conspecifics.

Habitat selection and lead poisoning

Two out of our eight study animals died from lead poisoning whilst the GPS transmitters were active. Although the eagles were satellite tracked, the exact food items which caused these lead intoxication were impossible to reliably identify in the field after their death. The main source of lead poisoning in white-tailed eagles are carcasses or gut piles of shot game interspersed with lead bullet fragments or lead shot (Kenntner *et al.* 2001, Krone *et al.* 2003, Langgemach *et al.* 2006). These food sources are available to wildlife in all habitat types where people hunt, mainly in forest, grassland and arable land. 'Forest' and 'grassland' were used in proportion to their availability but 'arable land' was avoided by the study animals. If 'forest' and 'grassland' would not occasionally offer some attractive food we would expect patterns of avoidance for these habitats as well. This suggests that eagles used these two habitats on a regular basis but not as intensively as open water bodies or riparian vegetation. This would be consistent with the lesser importance of carrion compared to fish and waterfowl in the diet of the eagles (Struwe-Juhl 1996b, Nadjafzadeh *et al.* submitted).

The eagles were located significantly further away from water bodies during winter than summer. Since the availability of fish and waterfowl is reduced in winter (Nadjafzadeh *et al.* submitted), the eagles might forage for alternative food resources such as carcasses or gut piles in larger areas, making them more threatened to lead poisoning. This is reflected by the fact that most lead-poisoned eagles are found in winter whereas only very few cases of lead intoxication were observed during summer (Krone *et al.* 2009). The two lead-poisoned study animals were also found during winter months (November and January).

Implications for management

Central Europe is still in the process of ongoing extensive habitat conversion. Conservation efforts regarding white-tailed eagles should focus on the preservation of environmental features identified as essentially important to eagles such as riparian vegetation, lakes and rivers. Areas close to water bodies are often particularly interesting for tourism and industry. As white-tailed eagles clearly avoid the close proximity of settlements, the development of new urban or touristic infrastructure at shorelines used by eagles should be strictly limited. This includes roads which also should not be built too close to the shore. Recreational activities in form of intensive water sports or fishing can distinctively degrade the quality of a lake or river as a hunting ground for white-tailed eagles and therefore need to be evaluated in terms of their impact in closer detail. Nest site characteristics were not an explicit topic of this paper but as we know from our own observations and the literature (Oehme 1961, Struwe-Juhl 1996a), large old trees in an undisturbed setting are essential for nest building and successful breeding. Protection of these sites is thus a prerequisite for successful white-tailed eagle management. We suggest that the management implications of this study can be widely extrapolated to other German and central European white-tailed eagle populations that face similar climatic and environmental conditions.

Acknowledgements

This study was financed by the Federal Ministry of Education and Research of Germany (BMBF) and administrated by the Project Management Juelich (PtJ). We thank Dr. Wolfgang Mewes, Dr. Wolfgang Neubauer, Jörg Gast and all staff members of the nature park *Nossentiner/ Schwinzer Heide* for their aid and logistic support. Accomodation in the field was kindly provided by the Reepsholt Foundation. We also thank Dr. Johannes Prüter for his advice and administrative help in the biosphere reserve *Niedersächische Elbtalaue.* We are grateful for the support of many hunters, fishermen and landowners in our study sites who provided baits and hides or gave permission to using their properties for eagle capturing.

References

Aebischer NJ, Robertson PA, and Kenward RE (1993). Compositional analysis of habitat use from animal radio-tracking data. *Ecology* 74: 1313-1325.

Alldredge JR and Griswold J (2006). Design and analysis of resource selection studies for categorical resource variables. *Journal of Wildlife Management* 70: 337-346.

Alldredge JR, Thomas DL, and McDonald L (1998). Survey and comparison of methods for study of resource selection. *Journal of Agricultural, Biological, and Environmental Statistics* 3: 237-253.

Benson JF and Chamberlain MJ (2007). Space use and habitat selection by female Louisiana black bears in the Tensas River Basin of Louisiana. *Journal of Wildlife Management* 71: 117-126.

Bloom P, Scott J, Pattee O, Smith M, Meyburg B, and Chancellor R (1989). Lead contamination of golden eagles *Aquila chrysaetos* within the range of the California condor *Gymnogyps californianus*. In: Meyburg BU and Chancellor RD (eds.) *Raptors in the modern world*, pp. 481-482. WWGBP, London, UK.

Boyce MS, Vernier PR, Nieslen SE, and Schmiegelow FKA (2002). Evaluating resource selection functions. *Ecological Modelling* 157: 281-300.

Cade TJ (2007). Exposure of California condors to lead from spent ammunition. *Journal of Wildlife Management* 71: 2125-2133.

Conner LM, Smith MD, and Burger LW (2003). A comparison of distance-based and classification-based analyses of habitat use. *Ecology* 84: 526-531.

del Hoyo J, Elliot A, and Sargatal J (1994) Family Accipitridae (hawks and eagles). In: de Hoyo J, Elliot A, Sargatal J (eds.) *Handbook of the birds of the world: vultures to guineafowl*, pp. 52-105. Lynx Edicions, Barcelona, Spain.

Dickson BG and Beier P (2002). Home-range and habitat selection by adult cougars in southern california. *The Journal of Wildlife Management* 66: 1235-1245.

Fischer W (1982). Die Seeadler. A. Ziemsen Verlag, Wittenberg Lutherstadt, Germany.

Fisher IJ, Pain DJ and Thomas VG (2006). A review of lead poisoning from ammunition sources in terrestrial birds. *Biological Conservation* 131: 421-432.

Folkestad AO (2003). Nest site selection and reproduction in the white-tailed sea eagle in Møre & Romsdal County, Western Norway in relation to human activity. In: Helander B, Marquiss M, and Bowerman, W. (eds.) *Sea Eagle 2000. Proceedings from the International Sea Eagles Conference in Bjorko, Sweden, 13-17 September 2000,* pp. 365-370, Swedish Society for Nature Conservation/ SNF, Stockholm, Sweden.

Gaines WL, Lyons AL, Lehmkuhl JF, and Raedeke KJ (2005). Landscape evaluation of female black bear habitat effectiveness and capability in the North Cascades, Washington. *Biological Conservation* 125: 411-425.

Gitzen RA, Millspaugh JJ, and Kernohan BJ (2006). Bandwith selection for fixed-kernel analysis of animal utilization distributions. *The Journal of Wildlife Management* 70: 1334-1344.

Harrison S and Bruna E (1999). Habitat fragmentation and large-scale conservation: what do we know for sure? *Ecography* 22: 225-232.

Hauff P (2008). Seeadler erobert weiteres Terrain. *Nationalatlas aktuell 1 (01/2008)* URL: http://NADaktuell.ifl-leipzig.de. Leipzig, Lcibniz-Institut für Länderkunde (ifL).

Hauff P and Mizera T (2006). Verbreitung und Dichte des Seeadlers *Haliaeetus albicilla* in Deutschland und Polen: eine aktuelle Atlas Karte. *Vogelwelt* 44: 134-136.

Hauff P (2003). Sea eagles in Germany and their population growth in the 20th century. In: Helander B, Marquiss M, and Bowerman W (eds.) *Sea Eagle 2000. Proceedings from the International Sea Eagles Conference in Bjorko, Sweden, 13-17 September 2000,* pp. 211-218. Swedish Society for Nature Conservation/ SNF, Stockholm, Sweden.

Hauff P (2001). Horste und Horstbäume des Seeadlers *Haliaeetus albicilla* in Mecklenburg-Vorpommern. *Berichte Vogelwarte Hiddensee* 16: 159-169.

Helander B and Sternberg T (2002). Action plan for the conservation of white-tailed sea eagles (*Haliaeetus albicilla*). 41 pp., Birdlife International, Strasbourg, France.

Hernandez M and Margalida A (2009). Assessing the risk of lead exposure for the conservation of the endangered Pyrenean bearded vulture (*Gypaetus barbatus*) population. *Environmental Research* 109: 837-842.

Hirzel AH, Hausser T, Chessel D, and Perrin N. (2002). Ecological-niche factor analysis: how to compute habitat-suitability maps without absence data? *Ecology* 83: 2027-2036.

Hunt WG, Burnham W, Parish CN, Burnham KK, Mutch B and Oaks JL (2006). Bullet fragments in deer remains: Implications for lead exposure in avian scavengers. *Wildlife Society Bulletin* 34: 167-170.

Johnson DH (1980). The comparison of usage and availability measurements for evaluating resource preference. *Ecology* 61: 65-71.

Kenntner N, Tataruch F, and Krone O (2001). Heavy metals in soft tissue of white-tailed eagles found dead or moribund in Germany and Austria from 1993 to 2000. *Environmental Toxicology and Chemistry* 20: 1831-1837.

Kim EY, Goto R, Iwata H, Masuda Y, Tanabe S and Fujita S (1999). Preliminary survey of lead poisoning of Steller's sea eagle (*Haliaeetus pelagicus*) and white-tailed sea eagle (*Haliaeetus albicilla*) in Hokkaido, Japan. *Environmental Toxicology and Chemistry* 18: 448-451.

Krone O, Kenntner N, and Tataruch F (2009). Gefährdungsursachen des Seeadlers (*Haliaeetus albicilla* L. 1758). *Denisia* 27: 139-146.

Krone O, Langgemach T, Sömmer P, and Kenntner N (2003). Causes of mortality in white-tailed sea eagles from Germany. In: Helander B, Marquiss M, and Bowerman W (eds.) *Sea Eagle 2000. Proceedings from the International Sea Eagles Conference in Bjorko, Sweden, 13-17 September 2000*, pp 211-218. Swedish Society for Nature Conservation/ SNF, Stockholm, Sweden.

Langgemach T, Kenntner N, Krone O, Müller K, and Sömmer P (2006). Anmerkungen zur Bleivergiftung von Seeadlern (*Haliaeetus albicilla*). *Natur und Landschaft* 81: 320-326.

Lele SR (2009). A new method for estimation of resource selection probability function. *Journal of Wildlife Management* 73: 122-127.

Lima SL (1998). Nonlethal effects in the ecology of predator-prey interactions. *Bioscience* 48: 25-34.

Manly BFJ, McDonald L, Thomas DL, McDonald T, and Erickson WP (2002). Resource selection by animals. Statistical designs and analysis for field studies. 2 edition. Kluwer Academic Publishers, Dordrecht, The Netherlands.

Mao JS, Boyce MS, Smith DW, Singer FS, Vales DJ, Vore JM, and Merril EH (2005). Habitat selection by elk before and after wolf reintroduction in Yellowstone National Park. *Journal of Wildlife Management* 69: 1691-1707.

Marzluff JM, Knick ST, Vekasy MS, Schueck LS, and Zarriello TJ (1997). Spatial use and habitat selection of Golden Eagles in Southwestern Idaho. *The Auk* 114: 673-687.

McClean SA, Rumble MA, King RM, and Baker WL (1998). Evaluation of resource selection methods with different definitions of availability. *The Journal of Wildlife Management* 62: 793-801.

Menzel JM, Ford MW, Menzel MA, Carter TC, Gardner JE, Gardner JD, and Hofmann JE (2005). Summer habitat use and home-range analysis of the endangered indiana bat. *The Journal of Wildlife Management* 69: 430-436.

Morosinotto C, Thompson RL and Korpimäki E (2010). Habitat selection as an antipredator behaviour in a multi-predator landscape: all enemies are not equal. *Journal of Animal Ecology* 79: 327-333.

Neu CW, Byers RC, and Peek JM (1974). A technique for analysis of utilization-availability data. *Journal of Wildlife Management* 38: 541-545.

Newton I (1979). Population ecology of raptors. T. & A. D. Poyser, London, UK.

Oehme G (1961). Die Bestandsentwicklung des Seeadlers, *Haliaeetus albicilla* (L.), in Deutschland mit Untersuchungen zur Wahl der Brutbiotope. In: Schildmacher H (ed.) *Beiträge zur Kenntnis deutscher Vögel,* pp. 1-61. Gustav Fischer Verlag, Jena, Germany

Oehme G (1975). Zur Ernährungsbiologie des Seeadlers (*Haliaeetus albicilla*), unter besonderer Berücksichtigung der Populationen in den drei Nordbezirken der Deutschen Demokratischen Republik. PhD thesis, Universität Greifswald, Germany

Oppel S, Schaefer HM, Schmidt V and Schröder B (2004). Habitat selection by the pale-headed brush-finch (*Atlapetes pallidiceps*) in southern Ecuador: implications for conservation. *Biological Conservation* 118: 33-40.

Palomares F, Delibes M, Ferreras P, Fedriani JM, Calzada J, and Revilla E (2000). Iberian lynx in a fragmented landscape: predispersal, dispersal, and postdispersal habitats. *Conservation Biology* 14: 809-818.

Powell BF and Steidl RJ (2002). Habitat selection by riparian songbirds breeding in southern Arizona. *Journal of Wildlife Management* 66: 1096-1103.

Schadt SA, Revilla E, Wiegand T, Knauer F, Kaczensky P, Breitenmoser U, Bufka L, Cerveny J, Koubek P, Huber T, Stanisa C, and Trepl L (2002). Assessing the suitability of central European landscapes for the reintroduction of Eurasian lynx. *Journal of Applied Ecology* 39: 189-203.

Schröder W (1998). Challenges to wildlife management and conservation in Europe. *Wildlife Society Bulletin* 26: 921-926.

Sinclair ARE, Fryxell JM, and Caughley G (2006). Wildlife ecology, conservation, and management. 2nd edition. Blackwell Science, Oxford, UK.

Struwe-Juhl B (1996a). Untersuchungen zur Habitatausstattung von Seeadler-Lebensräumen in Schleswig-Holstein. Forschungsstelle Wildbiologie/ Staatliche Vogelschutzwarte Schleswig-Holstein, Kiel, Germany

Struwe-Juhl B (1996b). Brutbestand und Nahrungsökologie des Seeadlers *Haliaeetus albicilla* in Schleswig-Holstein mit Angaben zur Bestandsentwicklung in Deutschland. *Vogelwelt* 117: 341-343.

Sulawa J (2009). Impact of lead poisoning on the dynamics of a recovering population: a case study of the German white-tailed eagle (*Haliaeetus albicilla*) population. PhD thesis, Freie Universität Berlin, Germany.

Sulkava S, Tornberg R, and Koivusaari J (1997). Diet of the white-tailed eagle *Haliaeetus albicilla* in Finland. *Ornis Fennica* 74: 65-78.

Swihart RK and Slade NA (1997). On testing for independence of animal movements. *Journal of Agricultural, Biological, and Environmental Statistics* 2: 48-63.

Thomas DL and Taylor EJ (1990). Study designs and tests for comparing resource use and availability. *Journal of Wildlife Management* 54: 322-330.

Thomas DL and Taylor EJ (2006). Study designs and tests for comparing resource use and availability II. *The Journal of Wildlife Management* 70: 324-336.

Worton BJ (1989). Kernel methods for estimating the utilization distribution in home range studies. *Ecology* 70: 164-168.

CHAPTER 4

Predicting suitable sites for recolonisation by a recovering and expanding raptor population: white-tailed eagles in a central European landscape

Abstract

As a result of persistent human persecution, the European populations of white-tailed eagles (*Haliaeetus albicilla*) were widely extirpated or reduced to critical low numbers at the beginning of the 20th century. Owing to conservation efforts the species recovered in several countries, but is still absent from Western Europe. Today, the German eagle population forms the western border of the distribution and is an important source population for whole Europe. We built a habitat model in order to predict future recolonisation sites across Germany and to roughly estimate possible prospective population size based on habitat availability. We used presence-absence data collected in Mecklenburg-Western Pomerania to train several candidate models by using generalised linear models. Our final model revealed a positive association of eagle presence with the percentage cover of open water and nature reserves within each grid cell. Eagle presence was negatively associated with the proportion of settlements. In total, 21 % of Germany's surface was classified as suitable for the establishment of white-tailed eagle territories. These results are consistent with an asymptotic stable population size of 1139 breeding pairs as predicted by Sulawa et al. (2009) for 2040. This prediction is based on the assumption that habitat patches considered as suitable for white-tailed eagles by our model will be protected from major habitat conversions. If this holds true, then there is a good chance that this raptor will expand further into its former western European range within the coming decades.

Keywords: Habitat modelling, *Haliaeetus albicilla,* logistic regression, raptor conservation, species recovery, white-tailed eagle

Introduction

White-tailed eagles (*Haliaeetus albicilla*) were once widespread throughout the Palaearctic but have been eradicated from most western and southern European countries by persistent human persecution during the past centuries (Fischer 1982, del Hoyo *et al.* 1994). The ongoing recovery of the German white-tailed eagle population from the brink of extinction is a success story for conservation. Once crashed to around 15 breeding pairs in 1900, the

German population is currently estimated to comprise 570 breeding pairs (Hauff 2008, Hauff 2000). Important milestones for effective conservation included comprehensive legal protection starting in 1934 and the prohibition of the pesticide dichlorodiphenyl-trichloroethane (DDT) in the 1970s (Oehme 1961, Kollmann *et al.* 2002, Helander and Sternberg 2002). Today, the main anthropogenic threats to white-tailed eagles are lead intoxication, train accidents, electrocution and poisoning (Krone *et al.* 2003, Kenntner *et al.* 2001). Germany has a special responsibility for white-tailed eagle protection because eastern and northern Germany forms the western border of the current distribution and the centres of its north-eastern eagle population function as a source population for the whole of Western Europe (Fig. 1). White-tailed eagles are mainly confined to riverrine, coastal and lake habitats but may be occasionally found away from larger water bodies if there is suitable food available (Fischer 1982, Willgohs 1961). They mainly feed on fish, waterfowl and carrion with seasonally changing proportions of these food items (Oehme 1975, Nadjafzadeh *et al.* submitted).

Germany is an industrial country with a high human population density (231 people km^{-2}, Statistisches Bundesamt 2008), extensive agricultural areas and a dense traffic network and is therefore mainly characterised by anthropogenically formed landscapes. As a consequence, the conservation of any large predator with pronounced space and habitat requirements is a challenge. Similar to large mammalian carnivores (Linnell *et al.* 2000, Schröder 1998), the conflict between human land use interests and those of wildlife is also important for the conservation of some raptors. Especially large eagles and vultures are often seen as a threat to livestock or as competitors of hunters. Consequences can be a controversial reputation, acceptance problems and illegal persecution (Graham *et al.* 2005, Galbraith *et al.* 2003, Phillips and Blom 1988, Redpath *et al.* 2004).

Raptors have one advantage over mammalian predators or scavengers in that they are able to cover long distances more easily and quickly. Habitat connectivity or the presence of habitat corridors are therefore likely to be less important, and good habitat patches are expected to be detected faster and re-occupied more easily (Hirzel *et al.* 2004). Therefore, habitat maps predicting the location and distribution of suitable habitats and hence areas of a high probability of recolonisation are of special value for the conservation management of highly mobile predatory species such as the white-tailed eagle. By providing a predictive habitat model on a larger scale, we aim to identify future recolonisation sites of particular conservation concern. We therefore concentrated on (1) detecting the distribution of suitable habitat patches, (2) estimating the amount of suitable habitat available for white-tailed eagles and (3) discussing our results in the light of recent population viability analyses. As the development of the German eagle population is of great importance for the reoccupation of Western Europe, implications from our study are of interest beyond national borders.

Material & Methods

Generating habitat models

A broad variety of methods are used to quantify species-habitat relationships such as ecological niche factor analysis (ENFA), discriminant function analysis, classification trees, artificial neural networks, generalised linear models (GLM) or generalised additive models (Segurado and Araújo 2004, Pearce and Ferrier 2000, Lek and Guégan 1999, De'ath and Fabricius 2000, Hirzel *et al.* 2002, McCullough and Nelder 1989). Several of these methods are based on presence-absence approaches (e.g. GLM) whilst others require presence-only data (ENFA). It is commonly stated that the inclusion of reliable absence data improves model quality and that GLMs perform better than ENFA if absences are true and species are overabundant (Brotons *et al.* 2004, Hirzel *et al.* 2001, Olivier and Wotherspoon 2006).

Whereas presence data are often available, acquiring proper absence data is much more difficult. This problem is of special concern for species such as the white-tailed eagle which return into their original range after a severe decline: non-occupied habitats do not necessarily have to be unsuitable habitats as the animals may choose this habitat as recolonisation proceeds. We therefore decided to use eagle distribution data of the federal state of Mecklenburg-Western Pomerania which encompasses the core of the German white-tailed eagle population. Here, the eagle population can be regarded as fairly saturated; absence data are therefore likely to indicate true absence and be suitable for use in habitat models. We thus chose a generalised linear model (GLM) and the classical presence-absence approach as most appropriate for our purpose.

In the course of our long-term white-tailed eagle research project, we fitted eight adult eagles with global positioning system (GPS) transmitters to study ranging behaviour and habitat selection. These birds used home ranges with an average size of 15 km^2 (95 % minimum convex polygons, chapter 2, 3). We incorporated the information on home range size and habitat use gained by our GPS telemetry study into the modelling process to improve the value and predictive power of our habitat model.

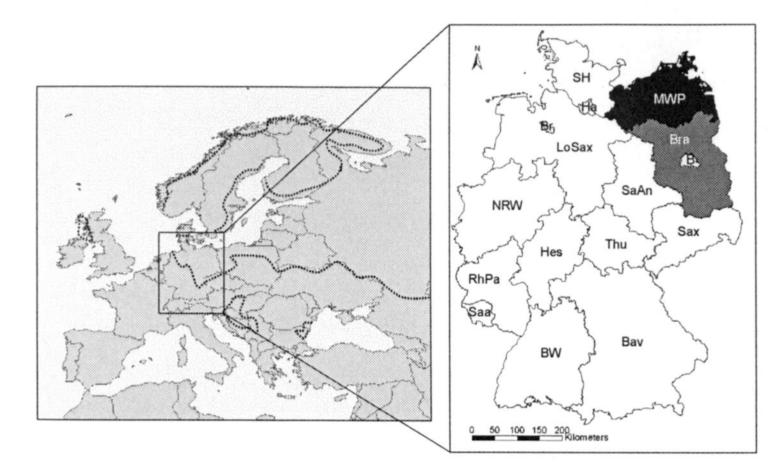

Fig. 1. European distribution of main white-tailed eagle populations (modified from Mebs, 2006) and Germany (inset). Dotted line: western margin of European eagle distribution. Inset, dark grey (MWP): the state of Mecklenburg-Western Pomerania from where the training data was taken. Inset, grey (Bra): the state of Brandenburg from where the evaluation data were taken

Study areas

For model building we used presence-absence data of the federal state Mecklenburg-Western Pomerania (MWP, Fig. 1) provided by the state office of environment (LUNG Güstrow). MWP is characterised by hilly lowlands, little industry, extensive pastures and some arable land, large managed forests and a rather low human population density (74 inhabitants km^{-2}) compared to the rest of Germany. The Baltic Sea forms a natural border in the North; the central inland is pervaded by the many water bodies of the Mecklenburg Lake District. Today, large parts of the state are protected to at least some extent, comprising 400 strict nature reserves, seven nature parks and three national parks. Owing to these favourable habitat conditions, the Mecklenburg Lake District was the place where the last few German white-tailed eagles survived persecution and from where recolonisation started (Hauff 1998, Hauff 2000).

For model evaluation we used data from the federal state of Brandenburg (Bra, Fig. 1). This state is located south of MWP, encompasses large tracts of forest and numerous lakes and also hosts a high number of white-tailed eagles. Model extrapolation and habitat mapping was performed over the entire area of Germany.

Distribution data

Resolution of records of white-tailed eagle presence in MWP was one quarter of a 1:25,000 plane survey sheet, resulting in an accuracy of approximately 5,400 m. Neither exact nest positions nor data with a higher resolution were available because of protection regulations. The evaluation data for Brandenburg were provided by the state office of environment in Potsdam and came in a lower resolution of one whole 1:25,000 plane survey sheet with an edge length of around 11,000 m.

GIS coverage

First, we built a grid using an edge length of 5,400 m, thereby reflecting the resolution of the distribution data. This resulted in 130 plots with eagle presence and 221 plots without eagle records within MWP. As one grid cell of 29.2 km^2 size could easily host one complete 16 km^2 eagle territory, we considered that the risk for data overlap and thus spatial autocorrelation effects was low and did not reduce this original data set. The Baltic coast region and a two-cell inland buffer were excluded from the data because there would have otherwise been a risk of overestimating the impact of water surface on eagle presence or absence during model building. For model extrapolation over the whole area of Germany, the coastal regions were included again.

We used CORINE land cover data (European Topic Center on Land Cover, Environment Satellite Data Center, Kiruna, Sweden) and ATKIS® habitat classification data (Basis-DLM and DLM250, Geodatenzentrum, Bundesamt für Kartographie und Geodäsie, Frankfurt/ Main). Whereas CORINE data came in a raster format with resolution of 100m, the fine-scaled ATKIS® data were vector based. The habitat types ‘forest’, ‘grassland’, ‘arable land’ and ‘settlement’ were taken from the CORINE data set; the habitat types ‘open water’ and ‘traffic’ were extracted from the ATKIS® data. Furthermore, a layer containing strict nature reserves and military training areas was derived from the ATKIS® data set (‘nature reserves’). The open water and the nature reserve layer were subsequently converted to raster format using a 100 m cell size to produce a consistent data set in line with the CORINE layers.

Model selection

We *a priori* compiled several hypotheses on eagle habitat preferences and compared all resulting model versions using an information-theoretic approach (Johnson and Omland 2004, Rushton *et al.* 2004). The seven habitat variables extracted from every cell and assessed to be important for eagle presence were percentage of ‘forest’, ‘grassland’, ‘arable land’, 'settlements', 'open water', 'nature reserves' and total length of 'traffic' routes (Table 1). The habitat types 'open water' (fish, waterfowl), 'forest', 'grassland', and 'arable land' (small mammals, carcasses, gut piles left by human hunters) were regarded as predictor variables related to foraging and food acquisition by eagles. We considered the variables 'settlements', 'traffic', and 'nature reserves' to represent the extent of human presence. As white-tailed eagles in Germany need undisturbed breeding sites, these three variables were hypothesised to be connected to nesting, in particular the variable 'nature reserve'.

The percentage cover of all variables within presence and absence cells was compared by a t-test. To avoid multicollinearity between explanatory variables, we calculated a correlation matrix of all seven predictor variables using Pearson's correlation coefficient. If the coefficient exceeded 0.5, the variable with the lower biological meaning was removed from the data set (Sachot and Perrin 2004). Based on the hypotheses that food availability and nesting sites are the predominant habitat requirements, the remaining explanatory variables were used to generate 14 candidate models, in which the importance of food and nesting constraints were varied (Table 2). These candidate models were fit to the training data set using generalised linear models with a logit link function and binomial error distribution. All candidate models were compared using the Akaike Information Criterion (Akaike 1973).

Model evaluation

For a threshold-independent measure of model accuracy we applied receiver-operating characteristics (ROC, Metz, 1978; Pearce and Ferrier, 2000): the true positive proportion of presence classifications (sensitivity) of the final model was plotted against the false positive proportion of predicted absences (specificity) for the whole range of possible thresholds. The power of the model to successfully discriminate presence and absence of eagles was assessed by measuring the area under the resulting ROC curve (AUC) and testing deviation from a random model for which the AUC would be 0.5. With regard to discrimination capacity, an AUC of between 0.7 and 0.8 typically is considered acceptable, an AUC between 0.8 and 0.9 excellent and values above 0.9 as outstanding (Pearce and Ferrier 2000, Hosmer and Lemeshow 2000).

Finding the appropriate threshold or cut-off value is essential for model evaluation and extrapolation (Jiménez-Valverde and Lobo 2007). The cut-off values we adopted were (1) the optimum threshold P_{OPT} producing the maximum number of correct classifications and (2) the threshold value P_{ROC} minimising the difference between sensitivity and specificity obtained from the ROC curve. As final cut-off value employed for model extrapolation we chose the value between P_{OPT} and P_{ROC} (Jiménez-Valverde and Lobo 2007, Guisan and Zimmermann 2000).

Habitat mapping and validation

We created raster maps with a grain size of 5,400 m for all habitat variables in analysis covering whole Germany. Each 5,400 m cell of every raster map was assigned the value of its particular percentage habitat cover. The variable 'traffic' was scored as the total length of roads and railways within each 5,400 m cell. For calculating the habitat suitability map we used the Raster Calculator function within the Spatial Analyst extension for ArcGIS (V. 8.2., ESRI Inc., Redlands, USA) and allocated our habitat suitability function to the respective raster habitat maps. Subsequently, the Neighbourhood Statistics function within the Spatial Analyst was applied to the resulting map. Every cell value of the new map now represented the mean probability of use of the respective cell together with its directly neighbouring cells. By this method, single, isolated suitable cells within unsuitable landscapes were devalued, whereas clusters of suitable habitat especially precious for eagle expansion were upgraded. In spite of the lower accuracy of this smoothed map ('management map') regarding absence prediction, this map was included into this study as it shows main habitat fragmentation patterns and centres of future recolonisation much clearer than the very detailed habitat suitability map.

To validate our final model, we calculated the proportion of correct classifications by our habitat suitability map for independent white-tailed eagle distribution data from the federal state of Brandenburg. The cell size of the Brandenburg presence-absence data was four times the resolution of our habitat suitability map. Therefore, a presence classification was regarded as correct if at least one fourth (one 5,400 m pixel) within the 1:25,000 plane survey sheet had a presence probability above our cut-off value.

Statistical analyses

All statistics were computed in R version 2.3.0 (R Development Core Team 2006, Vienna, Austria) and SPSS version 16 (SPSS Inc., Chicago, USA). Means are presented ± standard errors. Significance threshold was set as 0.05; p-values are given two-tailed.

Results

Exploratory univariate analysis

The predictor variables 'traffic' and 'settlement' as well as 'forest' and 'arable land' were highly correlated, therefore one variable per pair was excluded (Table 1). We kept the variable 'settlement' although the t-test for 'traffic' was significant, because the eagles wearing GPS-transmitters showed stronger avoidance of settlements than for traffic (chapter 3). White-tailed eagles do not need large continuous forests for breeding as long as single massive trees for nesting are available. Nests are often found within relatively small but undisturbed forest patches or groves. The tracked eagles showed no distinctive pattern in terms of preference or avoidance of forests, whereas all study animals distinctively avoided arable land (chapter 3). Thus, the variable 'forest' was excluded from the data set.

The remaining five explanatory variables included in habitat modelling were percentage cover by 'arable land', 'grassland', 'open water', 'settlements' and 'nature reserves'. Cells occupied by white-tailed eagles in MWP were covered by 44.5 ± 2.1 % 'arable land' and thus contained a significantly lower proportion of this land cover type than unused cells (60.5 ± 1.3 % cover, Table 1). The effects of the predictors 'open water' and 'nature reserve' were strong as well: used sites contained on average 8.3 ± 1.2 % 'open water' surfaces and 6.0 ± 0.9 % 'nature reserves', whereas sites of eagle absence only had proportions of 1.0 ± 0.3 %

Table 1. Predictor variables relevant to eagle presence and absence, S.E.: standard error, *: given by % cover unless indicated otherwise, ***: $p \leq 0.001$, R: Pearson's correlation coefficient; RV: variables retained for logistic regression (✓)

Variable	Behavioural context	Mean cover ± S.E. for presence cells*	Mean cover ± S.E. for absence cells*	P-value of T-Test	R > 0.5	RV
Forest	Foraging, Nesting	28.5 ± 1.8	18.8 ± 1.0	***	Arable land	
Grassland	Foraging	14.8 ± 1.2	15.2 ± 0.6			✓
Arable land	Foraging	44.5 ± 2.1	60.5 ± 1.3	***	Forest	✓
Settlement	Nesting	2.9 ± 0.3	3.9 ± 0.4		Traffic	✓
Open water	Foraging	8.3 ± 1.2	1.0 ± 0.3	***		✓
Nature reserves	Nesting	6.0 ± 0.9	1.9 ± 0.5	***		✓
Traffic (m)	Nesting	17,660.9 ± 636.4	21,119.6 ± 496.7	***	Settlement	

'open water' and 1.9 ± 0.5 % 'nature reserves' (Table 1). Furthermore, there was a non-significant trend for the proportion of 'settlement' to be smaller within used sites than within sites without eagle presence. The variable 'grassland' was of little relevance: presence-absence sites differed by just 0.4 % coverage.

Model selection and evaluation

Two models emerged as likely candidates for the final model: *model 1* included 'open water', 'settlement' and 'arable land' as significant predictors: it held the lowest AIC (= 381.8) of all

Table 2. The set of tested candidate models corresponding to the different behavioural constraints. AIC:Akaike's Information Criterion, Δ AIC: difference in model AIC to the AIC of the null model; *, **, ***: P-value of the log-likelihood test ($p \leq 0.05, 0.01, 0.001$)

Candidate models	AIC	Δ AIC
Null model (intercept only)	464.7	
Foraging		
Open water***	398.0	66.7
Arable land***	421.1	43.6
Grassland	464.6	0.1
Open water*** + arable land***	387.4	77.3
Open water*** + grassland + arable land***	388.7	76
Nesting		
Settlement	462.3	2.4
Nature reserves***	446.6	18.1
Settlement + nature reserves***	447.0	17.7
Foraging + Nesting		
Open water*** + settlement	395.2	69.5
Open water*** + nature reserve*	393.9	70.8
Open water* + settlement* + nature reserve* (*model 2*)**	**391.9**	**72.8**
Open water*** + settlement* + nature reserve* + grassland	393.5	71.2
Open water* + settlement** + nature reserve + arable land*** (*model 1*)**	**381.8**	**82.9**
Open water*** + settlement** + nature reserve + arable land*** + grassland	382.8	81.9

tested models (Table 2). The other (*model 2*) included 'open water', 'settlement', and 'nature reserve' (AIC = 391.9, Tab. 2). Despite the lower AIC of *model 1*, we decided to select *model 2* as final model for two reasons: (1) the ROC curves calculated for both models had a larger AUC for *model 2* than for *model 1*, thus *model 2* had the better discriminative performance (Table 3); (2) the habitat suitability map generated for *model 1* fitted the training and evaluation data in north-eastern Germany well but produced unrealistic predictions for the rest of the country, especially for the low mountain ranges of central and southern Germany. We therefore had to find a compromise between AIC assessment and extrapolation abilities for the chosen final model. In this regard, *model 2* performed best by including the three predictor variables hypothesised to be most important for the specific behavioural constraints ofthe eagles: percentage 'open water' (foraging for main prey) as well as 'settlement' and 'nature reserve' (breeding context, Table 2 and 3). The ROC plot for this model had an AUC of 0.83, indicating very good discrimination ability (AUC of *model 1* = 0.80, Table 3).

The threshold value obtained from the ROC curve (sensitivity = specificity) was P_{ROC} = 0.29. In contrast, results of classification trials with different cut-off values were best when using a threshold of P_{OPT} = 0.35. We therefore used the final cut-off value of P = 0.32 as intermediate between P_{ROC} and P_{OPT} for the extrapolation of our model.

Table 3. Parameter coefficients of *model 1* and *model 2*, S.E.: standard error, AIC: Akaike's Information Criterion, AUC: area under the receiver-operating characteristics curve, ***: a p-value of the test statistic of 0.001

Model	Predictor variable	Coefficient	S.E.[a]	AIC[b]	AUC[c]
Model 1	Grassland			381.8	0.80***
	Arable land***	-0.025	0.006		
	Settlement**	-0.073	0.032		
	Open water***	0.167	0.039		
	Nature reserve				
	(Intercept)	0.652	0.415		
Model 2	Settlement*	-0.049	0.029	391.9	0.83***
	Open water***	0.196	0.039		
	Nature reserve*	0.033	0.015		
	(Intercept)	-0.984	0.169		

The predictive performance of our final habitat suitability map was high: the Mecklenburg training data had a proportion of 0.81 correctly predicted cells and the Brandenburg evaluation data reached a value of 0.80 (Table 4). Applying the model to Germany resulted in a proportion of 0.86 correctly predicted occurrences but here no assessment of absence data was possible.

Habitat suitability map

Areas of suitable habitat were mainly concentrated in north-eastern Germany where habitat patches were well connected and formed several large continuous blocks (Fig. 3). Other clusters of suitable cells were located in the southernmost part of the country close to the Alps and on the North Sea and Baltic Sea coast. The remaining suitable habitat was rather patchily distributed across the rest of Germany. Our habitat model predicted that more than three quarters (78.9 %) of Germany's land area is unsuitable for white-tailed eagles (Fig. 2). The two highest probability classes covered 3.6 % and 5.7 % of cells; the lowest class above the required threshold included 11.7 % of all cells. With respect to the classification accuracy of 0.8 as found for our evaluation data, in total 2,174 suitable cells were identified throughout Germany. The number of cells in the two highest classes of presence probability was 964.

Table 4. Classification table for the training and evaluation data sets, MWP: Mecklenburg-Western Pomerania; [1]: training data; [2]: evaluation data; Res: spatial resolution; N: number of cells; [3]:correctly classified cells

Area	Res [m]	N presences	% + presences[3]	N absences	% + absences[3]	% + total[3]
MWP[1]	5400	130	79.2	221	82.3	0.81
Brandenburg[2]	10800	94	89.4	71	67.6	0.80
Germany[2]	10800	350	85.7	-	-	-

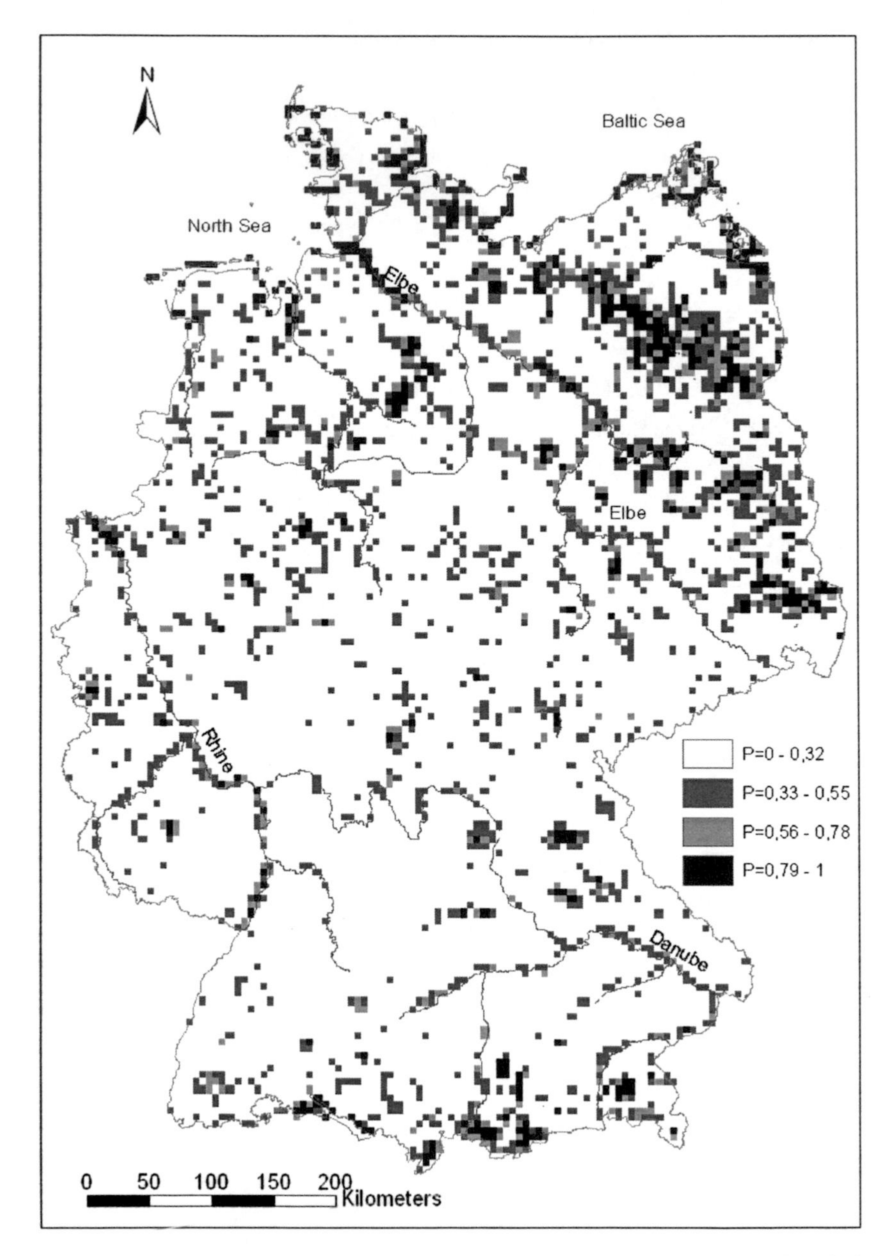

Fig. 2. Habitat suitability map showing predicted probabilities of occupancy by white-tailed eagles per cell, solid lines: main rivers

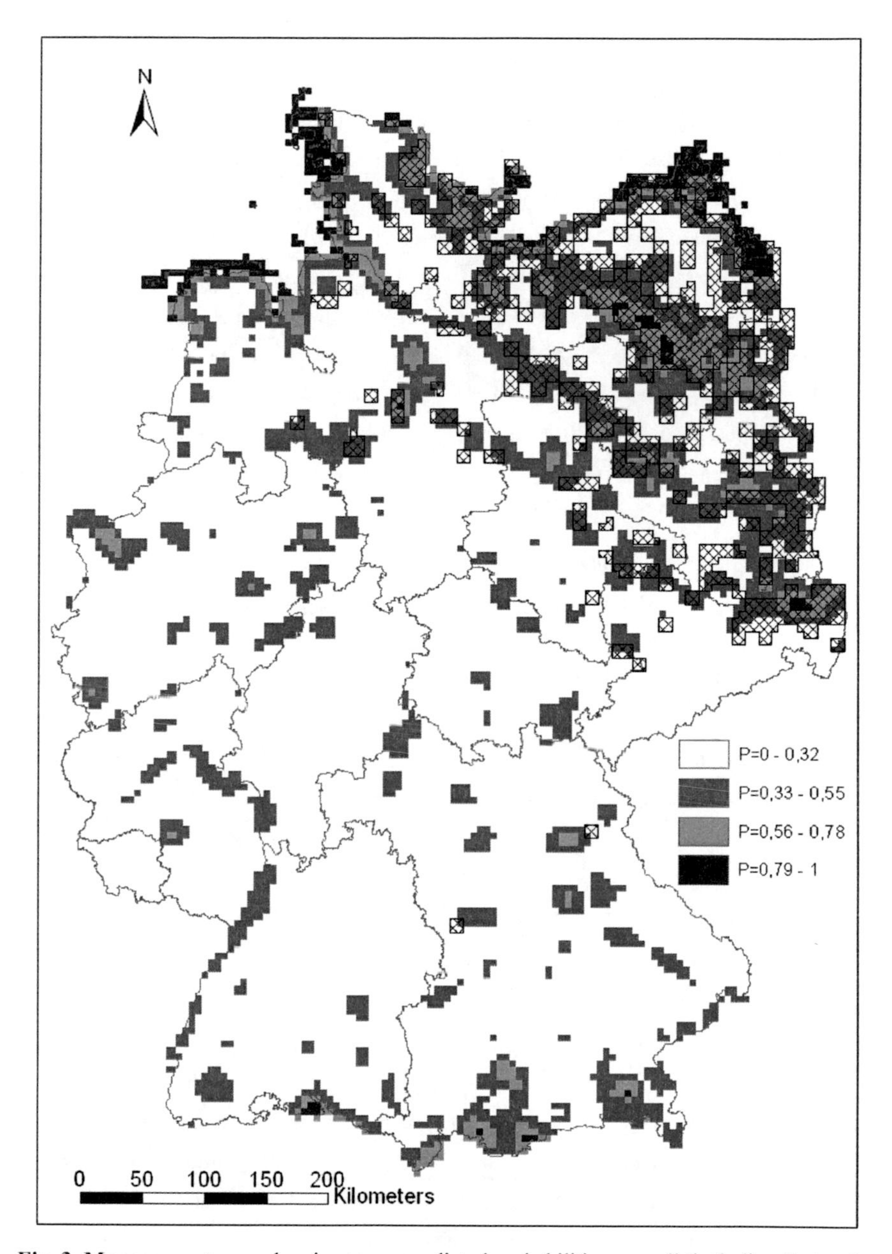

Fig. 3. Management map showing mean predicted probabilities per cell (including first order neighbour cells), dashed squares: all currently known cells with presence of white-tailed eagles (cell size 10,800m, Hauff 2008). Solid lines: boundaries of Federal states

Discussion

Model selection

When choosing between *model 1* and *model 2* to select the final model, we had to balance the trade-off between AIC assessment und ROC curve results. Both are widely used methods for estimating model quality but apply very different statistical approaches (Rushton *et al.* 2004, Klar *et al.* 2008, Guisan *et al.* 2007, McPherson *et al.* 2004, Osborne *et al.* 2001, Schadt *et al.* 2002). The AIC finds the most parsimonious model amongst a set of candidate models by penalising the maximum likelihood of the model for the number of model parameters (Akaike 1973, Burnham and Anderson 2002). In contrast, the ROC curve and the AUC are measures of the discriminatory power of a model (Pearce and Ferrier 2000). We decided to focus on discriminatory power rather than on parsimony because the overall aim of our study was to generate a habitat map that predicts eagle presence as well as absence as accurately as possible (Vaughan and Ormerod 2005).

An additional problem of *model 1* was the limited ability of accurate extrapolation. Since all candidate models were built using training data from lowland Germany, extrapolation over the whole range of Germany and its more mountainous and complex landscapes in the centre and south was a challenge. *Model 1* predicted extensive blocks of suitable habitat for all German low mountain ranges with almost no regard for the presence of lakes or rivers which is unlikely to be realistic. This was an effect of the predictor 'arable land' which was included in *model 1* and negatively associated with eagle presence. Thus, large areas with a low proportion of arable land were classified as good habitat and these areas were mostly forested. That resulted in an overestimation of the low mountain ranges which are predominantly covered by woods but have never been comprehensively occupied by white-tailed eagles.

Additionally, *model 1* included two predictors related to eagle foraging ('open water', 'arable land') but only one affecting nesting ('settlement'). As arable land is probably avoided because of scarce food supply to eagles, it is not a predictor directly enhancing foraging or nesting. In contrast, *model 2* included only explanatory variables with a direct positive effect on both foraging and nesting opportunities and focuses on landscape aspects associated with nesting: 'open water' provides the main food of eagles (Oehme 1975, Struwe-Juhl 1996) and they tend to breed in 'nature reserves' and areas with fewer 'settlements' and thus less human disturbance (Folkestad 2003). The habitat suitability map derived from *model 2* reflects this shift in focus and shows no obvious overestimation of areas without any larger water bodies or historic white-tailed eagle presence reports.

Constraints of our habitat model

Not all habitat features which might have influenced white-tailed eagle presence or absence were available as GIS layers for the whole of Germany, such as the degree of eutrophication of lakes and rivers and thus their potential food supply. Including the suitability of water bodies for eagle foraging might have improved model predictions, particularly within the two highest probability classes.

The predictions for the North Sea coast and its islands could be problematic. Parts of the region were classified as highly suitable despite the fact that many small North Sea islands are more or less bare and very exposed and lack possible breeding trees. There is a possibility that eagles might adapt to such conditions and nest on the ground as occasionally described from Scandinavia (Stjernberg 2003) but this is far from certain.

Many observed eagle occurrences classified as below the probability threshold (false negative classifications) were located in first order neighbouring cells of cells classified as suitable for white-tailed eagles. The eagle survey and thus all distribution data were based on nest site locations. An eagle nest and the respective waters bodies used for foraging can be several kilometres apart. If our model correctly identified the foraging grounds of an eagle pair but failed to predict the precise location of the breeding site, then the presence (nest) record would be found in first order neighbour cells of suitable cells.

Historical distribution and the predictions of recolonisation

Evidence in the literature for white-tailed eagle presence before 1850 is very scarce. Within many European cultures, eagles commonly appear in old emblems or flags symbolising power and dignity, making a former widespread distribution rather likely. However, these symbolic eagles may sometimes also refer to other species such as the golden eagle (*Aquila chrysaetos*). It can be assumed that white-tailed eagles originally occurred in suitable habitat across all of Europe (Fischer 1982, Mebs and Schmidt 2006, del Hoyo *et al.* 1994). In Germany in around 1850, the population was described to be mainly concentrated at the Baltic coast and its islands, the Mecklenburg Lake District and today's federal state of Brandenburg (Fischer 1982, Oehme 1961). South-east of the Elbe river, the number of white-tailed eagles was considerably lower if we believe older notes but they may still have been widespread. In southern Germany, single eagle pairs were reported to breed at the Danube and Rhine rivers (Oehme 1961), and the fact that 35 white-tailed eagles from Bavaria were stuffed in Augsburg between 1850 and 1870 alone suggests a much wider distribution of the species at least within Bavaria (Oehme 1961). It is quite possible that before the rise of the nature conservation movement, many occupied eagle territories were simply not noted because of a lack of interest in wildlife (Hauff 2000).

Our habitat map reflects the centres of historic distribution very well, since the largest continuous blocks of highly suitable habitat were generated for north-eastern Germany. South-east of the Elbe river, areas of suitable habitat are more patchily distributed but still cover a surface of considerable size. Alongside the Rhine and Danube rivers, reasonable amounts of good habitat are depicted which is consistent with historic eagle presence reports.

Present distribution and the predictions of recolonisation

The actual distribution of the re-expanding white-tailed eagle population is in good agreement with the predictions of our habitat model. The part of the population spreading into habitats outside the core population mainly inhabits areas with high predicted occupancy probabilities. Examples are the eagle territories recently established at the actual distribution margins in Lower Saxony, Saxony, Thuringia and Bavaria (Hauff 2008), which were correctly predicted by our model in most cases. In southern Germany, the predicted probabilities of occupancy by white-tailed eagles are high especially around the larger lakes (Chiemsee, Walchensee, Forggensee, Starnberger See, Ammersee, Bodensee). There have already been occasional sightings of white-tailed eagles at the Chiemsee (Ulrich Brendel, pers. comm.), thus the recolonisation of the southern Bavarian lakes seems to be just a matter of time.

Interestingly, white-tailed eagles near the core population tend to occupy less optimal habitat than they could if they moved away larger distances. This pattern is obvious for Mecklenburg-Western Pomerania which has the highest eagle population density in Germany. White-tailed eagles are capable of flying 180 km day^{-1} during juvenile dispersal as recorded by GPS-telemetry (Marion Westphal, pers. comm.), so exploring and colonising highly suitable new habitat should be rather easy in theory. Why does this not happen? An explanation for the tendency to stay close to the core population might be a tendency to philopatry. For instance, a better access to mating partners in these regions could be of importance. Philopatry is a common phenomenon among birds and raptors (Greenwood 1980) and probably also affects white-tailed eagle behaviour. Nevertheless, white-tailed eagles are currently expanding their range in Germany and have already established hundreds of new territories. Thus the actual effect of philopatry on the recolonisation process remains to be determined.

Implications for conservation

The habitat suitability map provided in this paper constitutes a useful tool for the future management of the expanding white-tailed eagle population in Germany. A next step would be an extrapolation of our habitat model over other European countries with historical white-tailed eagle populations.

We estimated a total amount of 2,174 cells offering suitable habitat for white-tailed eagles in Germany (= 63,394 km^2). One cell had a surface of 29.2 km^2 and would thus theoretically be capable of encompassing two white-tailed eagle home ranges of 15 km^2. Even with a more conservative approach assuming home ranges of around 35 km^2 (Oehme 1975), one cell could still host one eagle pair territory. Sulawa *et al.* (2009) predicted a minimum total population size of 1130 white-tailed eagle pairs for Germany in the year 2040 based on demographic models. Their study showed that the population is not expected to grow much larger once this number is reached because of negative density dependence and other demographic effects. Habitat data were not included in their models. As our habitat model generated 2,174 suitable cells with each having space for one or two eagle territories, the number of 1130 breeding pairs appears to be the right order of magnitude. It is also consistent with the 974 cells within the two highest probability classes. Negative density dependence as described and assumed in Sulawa *et al.* (2009) might be mitigated if there is sufficient suitable habitat available for the increasing population. We conclude that current knowledge on the distribution and availability of highly suitable habitats does not impose a constraint that is likely to act more severely than the currently recognised demographic feedback processes. Final asymptotic population size is therefore likely to reach or moderately exceed 1130 breeding pairs.

The German white-tailed eagle population forms the western border of the current range of the species and is an important source population for Western Europe. The recolonisation of Western Europe therefore depends to a large extent on the protection and support of the species in Germany. Our habitat maps identify foci for future recolonisation which need our special attention. Conservation efforts should concentrate on the protection of riparian habitats and old forest stands from urban development and logging to preserve the key prerequisites for the establishment of future eagle territories. This also implies that centres of future recolonisation should be safeguarded when planning new wind farms or other spacious structures likely to lower habitat quality (Krone and Scharnweber 2003, Bright *et al.* 2008).

Acknowledgments

This study was financed by the Federal Ministry of Education and Research of Germany (BMBF) and administrated by the Projektträger Jülich (PtJ). We are grateful to all people involved in the ongoing white-tailed eagle monitoring program in Germany who contributed to the comprehensive eagle distribution data set this study is based on. Furthermore, we thank S. Kramer-Schadt and N. Klar for their input regarding GIS data processing.

References

Akaike H (1973). Information theory as an extension of the maximum likelihood principle. In: Petrov BN and Csaki F. (eds.). *Second International Symposium on Information Theory*. pp. 267-281, Akademiai Kiado, Budapest, Hungary.

Bright J, Langston R, Bullman R, Evans R, Gardner S, and Pearce-Higgins J (2008). Map of bird sensitivities to windfarms in Scotland: A tool to aid planning and conservation. *Biological Conservation* 141: 2342-2356.

Brotons L, Thuiller W, Araújo MB, and Hirzel AH (2004). Presence-absence versus presence-only modelling methods for predicting bird habitat suitability. *Ecography* 27: 437-448.

Burnham KP and Anderson DR (2002). Model selection and multimodel inference. Springer, New York, USA.

De'ath G and Fabricius KE (2000). Classification and regression trees: a powerful yet simple technique for ecological data analysis. *Ecology* 81: 3178-3192.

del Hoyo J, Elliot A, and Sargatal J (1994) Family Accipitridae (hawks and eagles). In: de Hoyo J, Elliot A, Sargatal J (eds.) *Handbook of the birds of the world: vultures to guineafowl*, pp. 52-105. Lynx Edicions, Barcelona, Spain.

Fischer W (1982). Die Seeadler. A. Ziemsen Verlag, Wittenberg Lutherstadt, Germany.

Folkestad AO (2003). Nest site selection and reproduction in the white-tailed sea eagle in Møre & Romsdal County, Western Norway in relation to human activity. In: Helander B, Marquiss M., and Bowerman W. (eds.). *Sea Eagle 2000. Proceedings from the International Sea Eagles Conference in Bjorko, Sweden, 13-17 September 2000,* pp. 365-370, Swedish Society for Nature Conservation/ SNF, Stockholm, Sweden.

Galbraith CA, Stroud DA, and Thompson DBA (2003). Towards resolving raptor-human conflicts. In: Thompson DBA, Redpath SM, Fielding AH, Marquiss M, and Galbraith CA (eds.). *Birds of prey in a changing environment*. pp. 527-535, The Stationery Office, Edinburgh, UK.

Graham K, Beckerman AP, and Thirgood S (2005). Human-predator-prey conflicts: ecological correlates, prey losses and patterns of management. *Biological Conservation* 122: 159-171.

Greenwood PJ (1980). Mating systems, philopatry and dispersal in birds and mammals. *Animal Behaviour* 28: 1140-1162.

Guisan A, Graham CH, Elith J, Huettmann F, and the NCEAS Species Distribution Modelling Group. (2007). Sensitivity of predictive species distribution models to change in grain size. *Diversity and Distributions* 13: 332-340.

Guisan A and Zimmermann NE (2000). Predictive habitat distribution models in ecology. *Ecological Modelling* 135: 147-186.

Hauff P (1998). Bestandsentwicklung des Seeadlers *Haliaeetus albicilla* in Deutschland seit 1980 mit einem Rückblick auf die vergangenen 100 Jahre. *Vogelwelt* 119: 47-63.

Hauff P (2008). Seeadler erobert weiteres Terrain. *Nationalatlas aktuell 1 (01/2008)* URL: http://NADaktuell.ifl-leipzig.de. Leipzig, Leibniz-Institut für Länderkunde (ifL).

Hauff P (2003). Sea eagles in Germany and their population growth in the 20th century. In: Helander B, Marquiss M, and Bowerman W. (eds.). *Sea Eagle 2000. Proceedings from the International Sea Eagles Conference in Bjorko, Sweden, 13-17 September 2000,* pp. 211-218. Swedish Society for Nature Conservation/ SNF, Stockholm, Sweden.

Helander B and Sternberg T (2002). Action plan for the conservation of white-tailed sea eagles (*Haliaeetus albicilla*). 41 pp., Birdlife International, Strasbourg, France.

Hirzel AH, Hausser T, Chessel D, and Perrin N (2002). Ecological-niche factor analysis: how to compute habitat-suitability maps without absence data? *Ecology* 83: 2027-2036.

Hirzel AH, Helfer V, and Metral F (2001). Assessing habitat-suitability models with a virtual species. *Ecological Modelling* 145: 111-121.

Hirzel AH, Posse B, Oggier PA, Crettenand Y, Glenz C, and Arlettaz R (2004). Ecological requirements of reintroduced species and the implications for release policy: the case of the bearded vulture. *Journal of Applied Ecology* 41: 1103-1116.

Hosmer DW and Lemeshow S (2000). Applied logistic regression. 2nd edition. Wiley Interscience, Chichester, UK.

Jiménez-Valverde A and Lobo JM (2007). Threshold criteria for conversion of probability of species presence to either-or presence-absence. *Acta Oecologica* 31: 361-369.

Johnson JB and Omland KS (2004). Model selection in ecology and evolution. *Trends in Ecology and Evolution* 19: 101-108.

Kenntner N, Tataruch F, and Krone O (2001). Heavy metals in soft tissue of white-tailed eagles found dead or moribund in Germany and Austria from 1993 to 2000. *Environmental Toxicology and Chemistry* 20: 1831-1837.

Klar N, Fernandez N, Kramer-Schadt S, Herrmann M, Trinzen M, Büttner I, and Niemitz C (2008). Habitat selection models for European wildcat conservation. *Biological Conservation* 141: 308-319.

Kollmann R, Neumann T, and Struwe-Juhl B (2002). Bestand und Schutz des Seeadlers (*Haliaeetus albicilla*) in Deutschland und seinen Nachbarländern. *Corax* 19: 1-14.

Krone O, Langgemach T, Sömmer P, and Kenntner N (2003). Causes of mortality in white-tailed sea eagles from Germany. In: Helander B, Marquiss M, Bowerman W (eds.). *Sea Eagle 2000. Proceedings from the International Sea Eagles Conference in Bjorko, Sweden, 13-17 September 2000,* pp. 211-218. Swedish Society for Nature Conservation/ SNF, Stockholm, Sweden.

Krone O and Scharnweber C (2003). Two white-tailed sea eagles (*Haliaeetus albicilla*) collide with wind generators in northern Germany. *Journal of Raptor Research* 37: 174-176.

Lek S and Guégan JF (1999). Artificial neural networks as a tool in ecological modelling, an introduction. *Ecological Modelling* 120: 65-73.

Linnell JDC, Swenson JE, and Andersen R (2000). Conservation of biodiversity in Scandinavian boreal forests: large carnivores as flagships, umbrellas, indicators, or keystones? *Biodiversity and Conservation* 9: 857-868.

McCullough P and Nelder JA (1989). Generalized linear models. 2nd edition. Chapman & Hall, London, UK.

McPherson JM, Jetz W, and Rogers DJ (2004). The effects of species' range sizes on the accuracy of distribution models: ecological phenomenon or statistical artefact? *Journal of Applied Ecology* 41: 811-823.

Mebs T and Schmidt D (2006). Die Greifvögel Europas, Nordafrikas und Vorderasiens. Franckh-Kosmos Verlags GmbH, Stuttgart, Germany.

Oehme G (1975). Zur Ernährungsbiologie des Seeadlers (*Haliaeetus albicilla*), unter besonderer Berücksichtigung der Populationen in den drei Nordbezirken der Deutschen Demokratischen Republik. PhD thesis, Universität Greifswald, Germany.

Oehme G (1961). Die Bestandsentwicklung des Seeadlers, *Haliaeetus albicilla* (L.), in Deutschland mit Untersuchungen zur Wahl der Brutbiotope. In: Schildmacher H (ed.) *Beiträge zur Kenntnis deutscher Vögel*, pp. 1-61. Gustav Fischer Verlag, Jena, Germany.

Olivier F and Wotherspoon SJ (2006). Modelling habitat selection using presence-only data: Case study of a colonial hollow nesting bird, the snow petrel. *Ecological Modelling* 195: 187-204.

Osborne PE, Alonso JC, and Bryant RG (2001). Modelling landscape-scale habitat use using GIS and remote sensing: a case study with great bustards. *Journal of Applied Ecology* 38: 458-471.

Pearce J and Ferrier S (2000). Evaluating the predictive performance of habitat models developed using logistic regression. *Ecological Modelling* 133: 225-245.

Phillips R and Blom FS (1988). Distribution and magnitude of eagle/livestock conflicts in the western United States. In: Crabb AC, Marsh RE (eds.) *Proceedings of the Thirteenth Vertebrate Pest Conference,* pp. 241-244, University of California, Davis, USA.

Redpath SM, Arroyo BE, Leckie FM, Bacon P, Bayfield N, Gutierrez RJ and Thirgood SJ (2004). Using decision modeling with stakeholders to reduce human-wildlife conflict: a raptor-grouse case study. *Conservation Biology* 18: 350-359.

Rushton SP, Ormerod SJ, and Kerby G (2004). New paradigms for modelling species distributions? *Journal of Applied Ecology* 41: 193-200.

Sachot S and Perrin N (2004). Capercaillie (*Tetrao urugallus*) in western Switzerland. Viability and management of an endangered grouse metapopulation. In: Akçakaya HR, Burgman MA, Kindvall O, Wood CC, Sjögren-Gulve P, Hatfield JS, and McCarthy MA (eds.) *Species conservation and management: case studies,* pp. 384-396, Oxford University Press, Oxford, UK.

Schadt SA, Revilla E, Wiegand T, Knauer F, Kaczensky P, Breitenmoser U, Bufka L, Cerveny J, Koubek P, Huber T, Stanisa C, and Trepl L (2002). Assessing the suitability of central European landscapes for the reintroduction of Eurasian lynx. *Journal of Applied Ecology* 39: 189-203.

Schröder W (1998). Challenges to wildlife management and conservation in Europe. *Wildlife Society Bulletin* 26: 921-926.

Segurado P and Araújo MB (2004). An evaluation of methods for modelling species distributions. *Journal of Biogeography* 31: 1555-1568.

Statistisches Bundesamt (2008). Statistisches Jahrbuch 2008. Wiesbaden.

Stjernberg T (2003). Protection of nesting areas of the white-tailed sea eagle in Finland. In: Helander B, Marquiss M, and Bowerman W (eds.) *Sea Eagle 2000. Proceedings from the International Sea Eagles Conference in Bjorko, Sweden, 13-17 September 2000,* pp. 355-363. Swedish Society for Nature Conservation/ SNF, Stockholm, Sweden.

Struwe-Juhl B (1996). Brutbestand und Nahrungsökologie des Seeadlers *Haliaeetus albicilla* in Schleswig-Holstein mit Angaben zur Bestandsentwicklung in Deutschland. *Vogelwelt* 117: 341-343.

Sulawa J, Robert A, Köppen U, Hauff P, and Krone O (2009). Recovery dynamics and viability of the white-tailed eagle (*Haliaeetus albicilla*) in Germany. *Biodiversity and Conservation.* DOI 10.10007/s10531-009-9705-4.

Vaughan IP and Ormerod SJ (2005). The continuing challenges of testing species distribution models. *Journal of Applied Ecology* 42: 720-730.

Willgohs JF (1961). The white-tailed eagle *Haliaeetus albicilla albicilla* (Linné) in Norway. *Årbok for Universitetet i Bergen. Math.-Naturv. Serie*, pp. 212, Norwegian Universities Press, Bergen, Oslo, Norway.

CHAPTER 5

General discussion

The topic of this dissertation was a detailed investigation of key issues related to the spatial ecology of white-tailed eagles. Knowledge on space and habitat use of a species generally belongs to the most important prerequisites for successful protection efforts. In order to provide this information for white-tailed eagle management, I intensively studied the size and shape of home ranges, the ranging behaviour and patterns of habitat selection of this species by means of GPS telemetry. Furthermore, a habitat model was built for Germany to assess the amount and distribution of habitat suitable for future white-tailed eagle occupation.

Telemetry

Telemetry is a widely used method in the fields of animal ecology and conservation science. Because of its somewhat invasive character and the need to capture animals to fit transmitters, a clear concept, good research questions and the capacity for careful data analysis are important preconditions for a responsible use of this technique. Classic radio-telemetry uses radio waves at very high (VHF) or ultra high (UHF) frequencies to directly locate radio-collared animals in the field by triangulation from at least three observer points using an antenna (Kenward 2001). This method is not only rather inaccurate but also very time-consuming and staff-intensive. The benefits of classic radio-telemetry are that transmitters are comparatively cheap and light.

For many applications, modern telemetry systems using ARGOS satellites or the global positioning system (GPS) have supplanted radio-telemetry during the last years because of their many advantages. Modern GPS receivers used in wildlife telemetry like those adopted in this study are typically accurate within a range of 10-20 m (Hulbert and French 2001, Rodgers 2001). Furthermore, GPS positions are taken even in the remotest parts of study sites no matter whether they are accessible to human observers or not. GPS transmitters save much time and personal effort as positions are recorded automatically and sent to an office computer or cell phone via the Global System for Mobile Communication (GSM) or the ARGOS satellite transmission network or stored in the transmitters for later download. Disadvantages of the system are that GPS transmitters are expensive and still heavier than radio-transmitters because of the larger battery demands of the GPS element. However, for most larger mammals and birds the advantages of GPS telemetry clearly outweigh the drawbacks.

The GPS transmitters used in this study were developed in consideration of the special demands of white-tailed eagle telemetry by Vectronic Aerospace (Berlin) in cooperation with the project leader Oliver Krone. The transmitter weight of 170g accounted for approximately 3 % of the eagles body mass. This value is commonly regarded as an acceptable transmitter weight for birds (Withey *et al.* 2001, Kenward 2001).

Most transmitters were programmed to receive one position per day. Due to the limitations of transmitter and battery weight, this schedule was the best compromise between transmitter life span and data collection efficiency. Furthermore, spatial as well as temporal autocorrelation of location data could thus be avoided (Swihart and Slade 1997, Thomas and Taylor 2006). Since locations were taken every 25 hours in a circulating schedule, positions were recorded during the whole course of daylight. Of course, not every single eagle movement could be recorded with this approach. However, white-tailed eagles tend to perch most of the day and frequent and rapid changes of location are not very typical as evident from our visual observations in the field. We recorded many extraterritorial forays and other irregular behaviour despite the wide schedule of GPS data collection. I therefore argue that our data were representative for the ranging patterns of our study animals.

The habitat model

Habitat models and habitat suitability maps are useful tools to identify the main parameters which influence the distribution as well as sites of suitable habitat for future colonisation of a species (Guisan and Zimmermann 2000, Rushton *et al.* 2004). If the typical home range size of the studied species is known, habitat suitability maps allow an estimation of the potential magnitude of population size in a certain area based on habitat availability (= carrying capacity, Schadt *et al.* 2002, Haines *et al.* 2006,). Our model predicted suitable habitat for a maximum of 2,174 white-tailed eagle pairs. If only the two highest classes of probability of occupancy are considered, a potential population size of at least 964 breeding pairs might be possible. Sulawa (2009) estimated that the growth of the German white-tailed eagle population would reach a plateau at around 1,139 breeding pairs in 2040 by using population models based on demographic parameters. No habitat suitability map can assess whether single massive trees and good nest sites are available for breeding at a small scale level in an area predicted as generally suitable for white-tailed eagle territory establishment. Thus, not every site classified as suitable by the map will indeed be occupied by a white-tailed eagle pair.

My habitat model yielded a rate of 80 % correct eagle presence and absence classifications for the evaluation area of Brandenburg which is regarded as very good in the literature (Pearce and Ferrier 2000). This was probably due to the high accuracy of the distribution data which the model was based on. Most of the recently established new territories at the western distribution margin were correctly predicted by the model. Before the period of heavy persecution, white-tailed eagles probably were a rather common species in landscapes comprising water bodies throughout Germany (Fischer 1982, Oehme 1961). The recovery of the population since the ban of DDT is impressive; today the number of white-tailed eagles has again reached a level of around 570 breeding pairs (Hauff 2008). Given that conservation efforts continue and the positive population trend continues throughout the next decades, a final stable population size of 1,000 to 1,500 breeding pairs might be a realistic prediction for the second half of the 21th century.

Implications for habitat management

Landscapes untouched by human influences are almost non-existent in Europe. The continent is a mosaic of many different countries and languages and is one of the most densely populated areas in the world. As a result, massive habitat degradation and fragmentation pose the main problems for current wildlife conservation in Europe (Schröder 1998). Many species cope to live in human-altered environments or even urban areas (Møller 2009, Baker *et al.* 2007) but others have more pronounced habitat or space demands incompatible with highly modified environments. This problem is of special concern to Germany, an industrial country with a human population density as high as 231 inhabitants km^{-1} (Statistisches Bundesamt 2008) and very intensive forms of agriculture and forestry. Patches of good habitat are often dissected by roads and thus further fragmented. Rare species moving through large territories such as wolves (*Canis lupus*) may be able to live and disperse in semi-natural landscapes in Europe if there is enough food available, but road mortality remains an important problem for such species (Eggermann 2009). For other wildlife with more special habitat requirements such as Eurasian lynx (*Lynx lynx*), Eurasian otters (*Lutra lutra*) or wildcats (*Felis sylvestris*), roads, agricultural land and urban areas constitute major obstacles which might prevent dispersal and genetic exchange between the often small, isolated populations and (re)colonisation of new suitable habitats (Kramer-Schadt *et al.* 2004, Klar *et al.* 2008, Hauer *et al.* 2002).

The main problem for the conservation of white-tailed eagles in Germany other than lead intoxication is the ongoing conversion of natural habitats. The telemetry study presented in this dissertation revealed that the eagles strongly preferred water bodies and riparian vegetation and avoided settlements, arable land and, to a lesser extent, traffic. During the past two decades, the water quality of many water bodies substantially improved, and this had a positive effect on fish and waterfowl populations and therefore on the food availability of the eagles. However, many lakes, rivers or coastal sites are popular destinations for recreational activities which can be problematic in areas where white-tailed eagles are present. If water bodies used for foraging by eagles become too crowded by swimmers, boats and anglers, the eagles may avoid these places and have difficulties in finding enough food. Shorelines and riparian habitats are often visited by hikers, cyclists and other people and are favourite sites for the construction of houses, tourist huts or industrial infrastructure. As water bodies and habitats close to water are essentially important for eagles in terms of food acquisition and perching, the human impact on these habitats should be officially regulated in areas where white-tailed eagles are present. The telemetry data showed that several eagles undertook extraterritorial movements of up to 40 km away from their respective home range centres. This indicates that at least some eagles use their environment not only at home range but also at landscape scale. Thus, efforts to protect key habitats should be extended to the larger surrounding of white-tailed eagle home ranges.

The nests of all tracked birds and those of many others known to us were located in rather undisturbed sites within forests or small forest patches. This is consistent with the literature (Oehme 1961, Fischer 1982). The nests of four out of the eight study animals were located in strict nature reserves where humans are not allowed to enter. Nevertheless, a tendency for a decreased sensitivity to humans regarding nest site selection was reported for several new white-tailed eagle territories established during the last decades (Altenkamp *et al.* 2007, Hauff 2001). The adherence to existing nest site protection regulations should be strictly controlled. Since white-tailed eagles may build several nests within one home range, stands of old massive trees should be preserved also outside the actual nest site of a territorial eagle pair.

The habitat model established using distribution data of white-tailed eagles in Mecklenburg-Western Pomerania revealed that the presence of eagles was positively associated with the surface of water bodies and nature reserves within each grid cell and negatively associated with the amount of settlements. This corroborates the importance of water bodies for foraging and of nature reserves for breeding. Furthermore, the model illustrated the negative effect of urban areas on white-tailed eagle occurrence. Habitat fragmentation and isolation do not affect white-tailed eagle expansion as much as the expansion of mammalian carnivores. Since road mortality plays no important role and white-tailed eagles are capable of covering long distances during dispersal (Hailer *et al.* 2007, Westphal 2007, Nygård *et al.* 2000) the colonisation of new suitable habitat should be rather unproblematic. However, if patches of suitable habitat are only small isolated islands within unsuitable landscapes, these sites are probably less likely to be occupied by white-tailed eagles. Furthermore, the access to possible mating partners is presumably an important factor which influences the establishment of a new eagle territory. If there are no other eagles and thus no potential mating partners around, then even a larger patch of suitable habitat is probably less attractive for the eagles.

The expansion of the recovering white-tailed eagle population can be effectively supported by the preservation of habitat prerequisites important for future eagle occupancy. This should be done in concordance with the recommendations for eagle habitat and nest site protection from this study which are described in the section above.

Implications concerning natural causes of death

The most important natural cause of death among examined white-tailed eagles are territorial fights (Krone *et al.* 2009b). One of the three study animals which died whilst the GPS transmitters were active was killed by an intruder at the age of 17 years (eagle "655"). Furthermore, eagles may die of infections, malformation and starvation (Krone *et al.* 2009b). For the three latter causes of death no conclusions from our telemetry data can be drawn. However, our data allow an estimation of home range overlap and territoriality which are issues related to mortality by conspecific encounters.

It was the common view that white-tailed eagles do not defend home range borders insistently against neighbours and often tolerate foreign adult birds (Oehme 1975, Willgohs 1961, Fischer 1982, Glutz von Blotzheim 1971). Occasional aggressive behaviour against intruding adult eagles was described to be limited to the breeding season (Fischer 1982, Willgohs 1961). These statements are inconsistent with the assumption that "predators of vertebrates must have the exclusive use of a sufficient hunting range to ensure a steady food supply" (del Hoyo *et al.* 1994). As raptors typically behave territorial and actively defend home ranges (del Hoyo *et al.* 1994), a rather similar pattern seemed likely for white-tailed eagles as well. Struwe-Juhl (1996) recorded several territorial fights when he studied 10 eagle

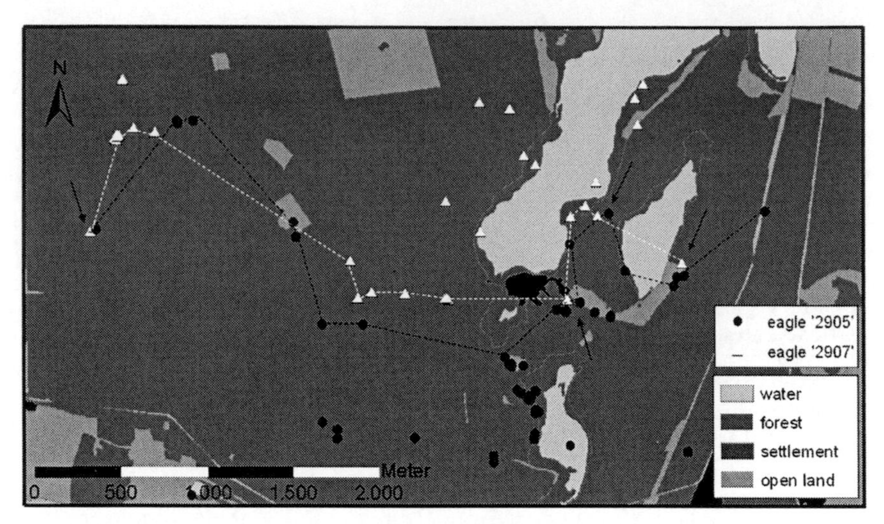

Fig. 1. Home range overlap of two eagles, northernmost ("2905") and southernmost ("2907") home range border, arrows: places of closest contact

pairs in *Schleswig-Holstein* during the breeding season. He concluded that the eagles defended home range borders against other adult birds and that home ranges did not overlap much. Krone (submitted) also stated that territorial behaviour plays an important role for white-tailed eagles throughout the year. All our study animals in the *Nossentiner/Schwinzer Heide* were directly neighboured to at least one other collared bird (see chapter 2, Fig. 1). Nevertheless, just two direct neighbours wore active transmitters within the same time span. These two birds spatially excluded each other rather distinctively (Fig. 1).

We had good evidence from long-term observations at our study site that home range use was quite stable over time. Thus, our data gives a useful estimate of home range overlap of all territories under investigation. Our data indicate clear territorial behaviour among white-tailed eagles. Home ranges overlapped little and defensive actions against neighbouring and foreign eagles were often observed. Such activities were mostly limited to intense vocalisations and pursuits; serious fights were rarely documented.

There is supposed to be an effect of population density on the intensity of defensive behaviour in raptors (del Hoyo *et al.* 1994). Our investigation mainly took place in the *Nossentiner/Schwinzer Heide* which hosts one of the highest densities of breeding pairs in Germany (Hauff and Mizera 2006). Therefore, our results do not have to be applicable to sparsely populated eagle habitats, although some kind of territorial behaviour is also expected to occur there (del Hoyo *et al.* 1994). Juveniles as well as immature eagles were in most observed cases tolerated by territorial birds which is in accordance with the literature (Fischer 1982, Glutz von Blotzheim 1971, Willgohs 1961).

Implications concerning anthropogenic causes of death

Anthropogenic causes of white-tailed eagle deaths include train accidents, electrocution, collisions with wire, poisoning, Hg-intoxication and collisions with wind power plants (Krone *et al.* 2009b, Krone and Scharnweber 2003). However, around 23 % of all white-tailed eagles found dead are documented to have died from lead poisoning (Krone *et al.* 2009b). Lead intoxication is not only the most common anthropogenic cause of death, but accounts for most white-tailed eagle deaths in general. In the following sections I will therefore concentrate on implications of this study concerning lead poisoning.

Cases of lead poisoning among the eight study animals

Two out of the three study animals which were found dead during the study period were proven to have died by lead intoxication. Both individuals were found in winter (November and January, respectively). One eagle had lead bullet fragments in the oesophagus (Krone *et al.* 2009a) and analyses of blood and liver samples of both birds revealed lethal concentrations of the heavy metal. Owing to GPS telemetry, the movements of these two female eagles during their last days could be roughly reconstructed.

One of the birds flew to a small island within its home range 13 days before it died and did not leave this island anymore (eagle “472”). Only very small-scale movements were recorded during this period, probably partly the result of varying GPS detection accuracy. The time of death could be determined owing to temperature and activity sensors implemented in the transmitter. Necropsis showed that the stomach and crop were empty and the bird had no fat tissue anymore (Krone *et al.* 2009a).

The demise of the second female was also a process that took around two weeks (eagle “2905”). This individual was a very young adult as evident from iris colouration and small black tips at some white tail feathers. 14 days before its death the bird was detected at a very localised place within its home range for five days in a row (Appendix, Fig. 2). When Oliver Krone went to this location, he found the eagle sitting apathetically in a tree just a few metres above him (Oliver Krone, pers. comm.). Then the eagle appeared to recover to some extent and changed location. It visited a camera site outside its home range on the following days which was set up by my colleague Mirjam Nadjafzadeh and fed on the (uncontaminated) carcass provided there as bait (Appendix, Fig. 2). Then the eagle flew to a small lake outside its home range and did not move for another four days until we found it dying. This is the best documented case of lead poisoning of a white-tailed eagle in the wild (see also Appendix, Fig. 2).

Six days before its first breakdown, the eagle was located at the same site in a forest within its home range for two successive days. Later on, remains of a wild boar carcass were found exactly at this site (Mirjam Nadjafzadeh, pers. comm., Appendix, Fig. 2). Of course, it is very difficult to identify the contaminated food item in the field that caused a lead intoxication. However, it is rather likely that the wild boar carcass was the source of lead poisoning in this case. This female was the third white-tailed eagle that died by lead poisoning within this respective eagle territory just between 2006 and 2009. Interestingly, the male which inhabited the neighbouring territory to the north as well as another study animal were 17 and 15 years old when we caught them as evident from their bands. Obviously, they were less heavily exposed to lead. It seems that lead sources are dispersed irregularly in the environment depending on local hunting practices and game abundances within the eagle territories. If small carnivores shot using lead ammunition or contaminated gut piles are left uncovered in the open landscape or even brought to special feeding sites, the risk for white-tailed eagles to get poisoned is much higher than if hunters remove such items.

Home range use and lead poisoning

With an average of 15 km^2, the home ranges of the study animals were rather small and the eagles clearly concentrated their activities on areas close to water bodies. In the winter months the home ranges were significantly larger than during summer. As the availability of fish and waterfowl is decreased during winter (Nadjafzadeh *et al.* submitted), the eagles probably use larger areas in search for alternative food such as carrion. This is in accordance with the fact that significantly more lead poisoned white-tailed eagles are found during winter whereas almost no cases of lead intoxication are observed in the summer months (Krone *et al.* 2009b).

Four eagles undertook only very rare and short excursions; for the four other birds more frequent and further movements outside their home ranges were recorded. Excursions were documented in winter as well as during the summer months. The average percentage of positions taken during such extraterritorial movements in relation to the total number of GPS positions was 5.9 %; or in other words, eagles were recorded within their respective home ranges in 94.1 % of GPS locations. Thus, the probability to feed on a lead contaminated food source is clearly highest within the home ranges of white-tailed eagles. The home ranges of all studied white-tailed eagles included grassland and/or forest where carcasses and gut piles could potentially be available to the eagles. It might occasionally happen that an eagle becomes poisoned by a lead source found during an extraterritorial movement but this is likely to be the exception rather than the rule. Furthermore, no eagle was recorded more than 38.6 km away from its home range centre and the mean distance covered during excursions was 14.4 km. Therefore, the area within which lead sources pose a risk for territorial white-tailed eagles can be limited to a rather small regional scale. The common view that territorial white-tailed eagles cover hundreds of kilometres and leave their territories for weeks and can thus be poisoned by lead sources very far away from their home ranges is not supported by our GPS data. Thus, foresters and hunters which are active within the extended home range areas of the eagles have the main responsibility for avoiding lead poisoning by spent ammunition. However, non-territorial white-tailed eagles and dispersing juveniles roam around through much larger areas (Westphal 2007), and several other species also suffer from lead exposure. Therefore, efforts to solve the lead problem should not be restricted only to areas where territorial white-tailed eagles are present.

Patterns of habitat selection with respect to lead poisoning

Human hunting activities are confined to certain habitats. Depending on local hunting practice, lead-contaminated gut piles and carrion are potentially available within these habitats. Probably most hunting is done in forests but hunters are also active in grassland and arable land. The eight white-tailed eagles tracked in the course of our study distinctively avoided arable land on population level as well as on individual level. This indicates that this habitat type was not very attractive for the eagles in terms of food. Probably, moribund game in search for shelter avoids open arable land. It is also possible that hunters regularly remove gut piles of shot game from cropland. Furthermore, the abundance of potential prey such as hares, rabbits or partridges is probably comparatively small on intensively used arable land.

On average, neither forest nor grassland were significantly preferred or avoided by the studied white-tailed eagles. However, on individual level most study animals showed a significant preference for either forest or grassland, depending on the abundance of both habitats in their respective activity ranges (Appendix, Table 1 + 2). These habitats were thus important to eagles, probably in terms of perching sites and food acquisition. This illustrates the need to mitigate threats by spent lead ammunition in habitats used by both human hunters and white-tailed eagles.

The excursions of the study animals were directed towards water bodies, forests as well as other habitat types and no clear seasonal pattern was found concerning habitats chosen on extraterritorial forays. Therefore, neither the search for carrion nor other single factors can be seen as the general driving force determining all extraterritorial movements. The motivation of eagles to undertake such movements is likely to be highly variable among individuals as well as single excursion events.

During winter months, the tracked white-tailed eagles were detected significantly further away from water bodies than during summer. This was probably a result of the decreased proportion of fish and waterfowl and an increased importance of carrion and other alternative food sources in the winter diet.

Implications of the habitat model regarding lead poisoning

The aim of the habitat model was to identify the main parameters influencing the distribution of white-tailed eagles and to generate a habitat suitability map which depicts suitable sites for future white-tailed eagle recolonisation. With regard to this species, lead poisoning will be a problem wherever white-tailed eagles recolonise landscapes and humans use lead ammunition for hunting. Thus, the habitat suitability map not only predicts suitable habitat for the establishment of new eagle territories, but also shows where conflicts regarding the lead poisoning problem will most likely arise in the future. In order to mitigate these conflicts, practises to avoid lead exposure to the environment will hopefully soon be common standard in areas where white-tailed eagles are present today. These practises should also be familiar to hunters and possibly already be implemented before the first eagles arrive at new sites. This includes the use of alternative lead-free ammunition or the removal of potentially lead-contaminated gut-piles and small carnivore carcasses from the countryside and a careful search for moribund game wounded by lead bullets.

If foresters and hunters are willing to support white-tailed eagles during winter, roadkills instead of potentially lead-contaminated gut piles could be provided with the permission of the responsible veterinary offices. Sulawa (2009) found a significant effect of lead poisoning on the growth rate of the recovering white-tailed eagle population. There is plenty of suitable habitat still available for the expanding population in Germany. Therefore, the further re-expansion of this raptor into its original range is not limited by habitat, but first of all by our efforts and willingness to protect and support this raptor species in Germany and throughout the rest of Europe.

Future directions

Tracking more white-tailed eagles living in riverine or coastal habitats would be a useful approach. As these landscapes are differently structured than lake habitats, white-tailed eagles might exhibit other ranging patterns. The home range and habitat use of the monitored male which inhabited a territory at the Elbe river ("2355") showed no distinctive deviation from the spatial use of the other eagles tracked in the *Nossentiner/Schwinzer Heide,* but this does not necessarily hold true for other riverine eagles. A larger number of eagles under investigation could help to further improve our knowledge on the spatial ecology and behaviour of this rarely studied species. With the development of new solar powered and lighter GPS transmitters, it will be possible to track individual white-tailed eagles much longer and allow researchers to gain detailed insight into the life history of this raptor.

The German population of white-tailed eagles currently forms the western border of the European distribution. If the positive population trend proceeds, an expansion of white-tailed eagles deeper into their former western and southern European range within the next decades is rather likely. An extrapolation of our habitat model over western European countries such as France or Spain could help to identify habitat suitable for white-tailed eagle territory establishment in these countries and to protect these sites from habitat degradation. By this means, the re-expansion of white-tailed eagles into these countries could be directly supported.

References

Altenkamp R, Stoewe D and Krone O (2007). Verlauf und Scheitern einer Brut des Seeadlers (*Haliaeetus albicilla*) in Berlin und Konsequenzen für den Schutz der Brutplätze. *Berliner ornithologischer Bericht* 17: 31-41.

Baker PJ, Dowding CV, Molony SE, White PC, and Harris S (2007). Activity patterns of urban red foxes (*Vulpes vulpes*) reduce the risk of traffic-induced mortality. *Behavioral Ecology* 18: 716-724.

del Hoyo J, Elliot A, and Sargatal J (1994) Family Accipitridae (hawks and eagles). In: de Hoyo J, Elliot A, Sargatal J (eds.) *Handbook of the birds of the world: vultures to guineafowl*, pp. 52-105. Lynx Edicions, Barcelona, Spain

Eggermann J (2009). The impact of habitat fragmentation by anthropogenic infrastructures on wolves *(Canis lupus)*. PhD thesis, University of Bochum, Germany.

Fischer W, (1982). Die Seeadler. A. Ziemsen Verlag, Wittenberg Lutherstadt, Germany.

Glutz von Blotzheim UN (1971). *Haliaeetus albicilla* (Linné 1758) - Seeadler. In: Glutz von Blotzheim UN., Bauer K, and Bezzel E (eds.) *Handbuch der Vögel Mitteleuropas*. pp. 169-203, Verlagsgesellschaft, Frankfurt/ Main, Germany.

Guisan A and Zimmermann NE (2000). Predictive habitat distribution models in ecology. *Ecological Modelling* 135: 147-186.

Hailer F, Helander B, Folkestad AO, Ganusevich S, Garstad S, Hauff P, Koren C, Masterov VB, Nygård T, Rudnick JA, Shiraki S, Skarphedinsson K, Volke V, Wille F, and Vilà C (2007). Phylogeography of the white-tailed eagle, a generalist with large dispersal capacity. *Journal of Biogeography* 34: 1193-1206.

Haines AM, Tewes ME, Laack LL, Horne JS, and Young JH (2006). A habitat-based population viability analysis for ocelots (*Leopardus pardalis*) in the United States. *Biological Conservation* 132: 424-436.

Hauer S, Ansorge H, and Zinke O (2002). Mortality patterns of otters (*Lutra lutra*) from eastern Germany. *Journal of Zoology* 256: 361-368.

Hauff P (2008). Seeadler erobert weiteres Terrain. *Nationalatlas aktuell 1 (01/2008)* URL: http://NADaktuell.ifl-leipzig.de. Leipzig, Leibniz-Institut für Länderkunde (ifL).

Hauff P (2001). Horste und Horstbäume des Seeadlers *Haliaeetus albicilla* in Mecklenburg-Vorpommern. *Berichte Vogelwarte Hiddensee* 16: 159-169.

Hauff P and Mizera T (2006). Verbreitung und Dichte des Seeadlers *Haliaeetus albicilla* in Deutschland und Polen: eine aktuelle Atlas Karte. *Vogelwelt* 44: 134-136.

Hulbert IAR and French J (2001). The accuracy of GPS for wildlife telemetry and habitat mapping. *Journal of Applied Ecology* 38: 869-878.

Kenward RE (2001). A manual for wildlife radio tagging. Academic Press, London, UK.

Klar N, Fernandez N, Kramer-Schadt S, Herrmann M, Trinzen M, Büttner I, and Niemitz C (2008). Habitat selection models for European wildcat conservation. *Biological Conservation* 141: 308-319.

Kramer-Schadt S, Revilla E, Wiegand T, and Breitenmoser U (2004). Fragmented landscapes, road mortality and patch connectivity: modelling influences on the dispersal of European lynx. *Journal of Applied Ecology* 41: 711-723.

Krone O, Berger A and Schulte R (2009a). Recording movement and activity pattern of a white-tailed sea eagle (*Haliaeetus albicilla*) by a GPS datalogger. *Journal of Ornithology* 150: 273-280.

Krone O, Kenntner N, and Tataruch F (2009b). Gefährdungsursachen des Seeadlers (*Haliaeetus albicilla* L. 1758). *Denisia* 27: 139-146.

Krone O and Scharnweber C (2003). Two white-tailed sea eagles (*Haliaeetus albicilla*) collide with wind generators in northern Germany. *Journal of Raptor Research* 37: 174-176.

Møller AP (2009). Successful city dwellers: a comparative study of the ecological characteristics of urban birds in the Western Palearctic. *Oecologia* 159: 849-858.

Nygård T, Kenward RE and Einvik K (2000). Radio telemetry studies of dispersal and survival in juvenile white-tailed sea eagles *Haliaeetus albicilla* in Norway. In: Meyburg B-U and Chancellor RD (eds.) *Raptors at Risk*. pp. 487-497, Hancock House, Surrey, Canada.

Oehme G (1961). Die Bestandsentwicklung des Seeadlers, *Haliaeetus albicilla* (L.), in Deutschland mit Untersuchungen zur Wahl der Brutbiotope. In: Schildmacher H (ed.) *Beiträge zur Kenntnis deutscher Vögel*, pp. 1-61, Gustav Fischer Verlag, Jena, Germany.

Oehme G (1975). Zur Ernährungsbiologie des Seeadlers (*Haliaeetus albicilla*), unter besonderer Berücksichtigung der Populationen in den drei Nordbezirken der Deutschen Demokratischen Republik. PhD thesis, Universität Greifswald, Germany.

Pearce J and Ferrier S (2000). Evaluating the predictive performance of habitat models developed using logistic regression. *Ecological Modelling* 133: 225-245.

Rodgers AR (2001). Recent telemetry technology. In: Millspaugh JJ and Marzluff JM (eds.) *Radio tracking and animal populations*. pp. 82-125, Academic Press, San Diego, USA.

Rushton SP, Ormerod SJ, and Kerby G (2004). New paradigms for modelling species distributions? *Journal of Applied Ecology* 41: 193-200.

Schadt SA, Revilla E, Wiegand T, Knauer F, Kaczensky P, Breitenmoser U, Bufka L, Cerveny J, Koubek P, Huber T, Stanisa C, and Trepl L (2002). Assessing the suitability of central European landscapes for the reintroduction of Eurasian lynx. *Journal of Applied Ecology* 39: 189-203.

Schröder W (1998). Challenges to wildlife management and conservation in Europe. *Wildlife Society Bulletin* 26: 921-926.

Statistisches Bundesamt (2008). Statistisches Jahrbuch 2008. Wiesbaden, Germany.

Struwe-Juhl B (1996). Untersuchungen zur Habitatausstattung von Seeadler-Lebensräumen in Schleswig-Holstein. Forschungsstelle Wildbiologie/ Staatliche Vogelschutzwarte Schleswig-Holstein, Kiel, Germany.

Sulawa J (2009). Impact of lead poisoning on the dynamics of a recovering population: a case study of the German white-tailed eagle (*Haliaeetus albicilla*) population. PhD thesis, Freie Universität Berlin, Germany.

Sulawa J, Robert A, Köppen U, Hauff P, and Krone O (2009). Recovery dynamics and viability of the white-tailed eagle (*Haliaeetus albicilla*) in Germany. *Biodiversity and Conservation* DOI 10.10007/s10531-009-9705-4.

Swihart RK and Slade NA (1997). On testing for independence of animal movements. *Journal of Agricultural, Biological, and Environmental Statistics* 2: 48-63.

Thomas DL and Taylor EJ (2006). Study designs and tests for comparing resource use and availability II. *The Journal of Wildlife Management* 70: 324-336.

Westphal M (2007). Bewegungen und Nutzungsräume junger Seeadler (*Haliaeetus albicilla*) in Deutschland. Diploma thesis, Freie Universität Berlin, Germany.

Willgohs JF (1961). The white-tailed eagle *Haliaeetus albicilla albicilla* (Linné) in Norway. *Årbok for Universitetet i Bergen. Math.-Naturv. Serie*. pp. 1-212, Norwegian Universities Press, Bergen, Oslo, Norway.

Withey JC, Bloxton TD, and Marzluff JM (2001). Effects of tagging and location error in wildlife radiotelemetry studies. In: Millspaugh JJ and Marzluff JM (eds.) *Radio tracking and animal populations*. pp. 45-75, Academic Press, London, UK.

SUMMARY

The still increasing rate of habitat degradation and fragmentation poses the main problem for nature conservation in central Europe today. Therefore, one of the most important prerequisites for successful protection efforts is knowledge on the space and habitat requirements of a species of interest. White-tailed eagles were almost eradicated in Germany at the beginning of the 20th century. Owing to extensive conservation efforts the German white-tailed eagle population recovered well in the course of the 20th century. Nevertheless, white-tailed eagles still suffer from various anthropogenic threats. The most important cause of death among white-tailed eagles in Germany are lead intoxications by lead bullet fragments and lead shot which are ingested together with food. The overall aim of this dissertation was to investigate the spatial use of adult, territorial white-tailed eagles and to discuss results with respect to habitat management and the lead poisoning problem.

Chapter 2 was focused on the examination of home range sizes, seasonal patters in home range use and the characterisation of extraterritorial forays. In the course of the project, seven white-tailed eagles were fitted with GPS (Global Positioning System) transmitters in the nature park *Nossentiner/Schwinzer Heide* and one male in the biosphere reserve *Niedersächsische Elbtalaue*. The eight study animals used relatively small home ranges of on average 15 km^2 size. The home ranges were significantly larger in winter; probably because of the decreased availability of fish and waterfowl during the winter months and a therefore increased search effort for alternative food sources such as gut piles or carrion.

Chapter 3 summarised the results of a detailed investigation of habitat selection patterns of the tracked white-tailed eagles at two spatial scales. Riparian vegetation and open water bodies were preferred regarding the habitat composition of the activity ranges as well as concerning GPS positions. In contrast, settlements and arable land were significantly avoided at both spatial scales. Forest, grassland and swamp/small waters were used in proportion to availability. These results demonstrate that next to the protection of nest sites the preservation of riparian habitats and water bodies from urban or industrial development and immoderate recreational activity is of essential importance for white-tailed eagle conservation. Furthermore, lead exposure should be mitigated in forest and/or grassland habitats used by both human hunters and white-tailed eagles.

Topic of chapter 4 was a habitat model which was developed using white-tailed eagle distribution data of the federal state of Mecklenburg-Western Pomerania. The logistic regression revealed a positive association of white-tailed eagle presence with the amount of open water and nature reserves within the examined grid cells, whereas settlements had a negative effect on eagle presence. The habitat suitability map illustrated that the expanding German white-tailed eagle population is currently not limited by the availability of suitable habitat.

Given that nest sites and important white-tailed eagle habitats will be protected and anthropogenic causes of death such as lead poisoning will be alleviated, the future perspectives for white-tailed eagles in Germany are positive. The replacement of lead ammunition by non-toxic alternatives can make an important contribution to the future progress of the white-tailed eagle population.

ZUSAMMENFASSUNG

Die zunehmende Degradierung und Fragmentierung verbliebener Naturräume stellt eines der Hauptprobleme für den Erhalt der Biodiversität in Mitteleuropa dar. Eine wichtige Voraussetzung für effektive Schutzmaßnahmen ist daher, grundsätzliche Raum- und Habitatansprüche einer zu schützenden Art zu kennen. Seeadler waren zu Beginn des 20. Jahrhundert in Deutschland beinahe ausgerottet. Dank intensiver Schutzbemühungen konnte sich die deutsche Seeadler-Population im Verlauf des 20. Jahrhundert wieder gut erholen; jedoch sind Seeadler weiterhin vielfältigen Gefahren ausgesetzt. Die wichtigste anthropogene Todesursache für Seeadler in Deutschland sind Bleivergiftungen durch Splitter bleihaltiger Büchsengeschosse sowie Bleischrote, welche mit der Nahrung aufgenommen werden. Das Ziel dieser Arbeit war, gründliche Untersuchungen zur Raumnutzung adulter, territorialer Seeadler anzustellen und die Ergebnisse in Bezug zum Habitatmanagement und zur Bleivergiftungsproblematik zu stellen.

Dafür wurden zunächst die Streifgebietsgrößen, saisonale Unterschiede in der Streifgebietsnutzung sowie außerterritoriale Streifzüge der Seeadler detailliert untersucht (Kapitel 2). Zu diesem Zweck wurden im Naturpark Nossentiner Heide und im Biosphärenreservat Niedersächsische Elbtalaue sieben bzw. ein Seeadler gefangen und mit GPS (Global Positioning System)-Sendern versehen. Die acht besenderten Seeadler nutzen relativ kleine Streifgebiete von durchschnittlich ca. 15 km^2. Die Streifgebiete waren im Winter signifikant größer als im Sommer. Das deutet darauf hin, dass die Adler aufgrund eines verringerten Angebots an Fischen und Wasservögeln im Winter auf der Suche nach Aas und Aufbrüchen weiter umherstreifen.

Weiterhin wurde die Habitatwahl der telemetrierten Seeadler auf zwei räumlichen Ebenen detailliert dargestellt (Kapitel 3). Ufervegetation und Wasserflächen wurden sowohl bei der Wahl des Aktionsraums als auch hinsichtlich der GPS-Lokalisationen präferiert, während Ackerland und Siedlungen signifikant gemieden wurden. Wald, Grünland und Sumpf/ Kleingewässer wurden entsprechend der Verfügbarkeit genutzt. Die Ergebnisse dieser Teilstudie zeigen, dass der Schutz ufernaher Bereiche und Wasserflächen vor Zersiedelung und übermäßiger Freizeitnutzung von essentieller Bedeutung für den Seeadler ist. Weiterhin sollte die Gefährdung durch Blei in Grünland- und/ oder Waldflächen, welche sowohl durch die Jagd als auch vom Seeadler genutzt werden, so weit wie möglich vermieden werden.

In einem nächsten Schritt wurde mit Hilfe einer logistischen Regression auf Grundlage von Seeadler-Verbreitungsdaten aus Mecklenburg-Vorpommern ein Habitatmodell erstellt

(Kapitel 4). Das Modell ergab, dass das Vorkommen von Seeadlern positiv mit dem flächenmäßigen Anteil an Wasserflächen und Naturschutzgebieten und negativ mit dem Anteil von Siedlungsflächen in den untersuchten Rasterzellen assoziiert ist. Die Habitateignungskarte verdeutlichte zudem, dass die sich ausbreitende Seeadler-Population in Deutschland momentan nicht durch die Verfügbarkeit geeigneter Habitat limitiert ist.

Wenn Brutplätze und für die Nahrungsbeschaffung wichtige Habitate der Seeadler auch in Zukunft geschützt werden und die Bedeutung anthropogener Todesursachen verringert werden kann, sind die Zukunftsaussichten für den Seeadler in Deutschland positiv. Der Gebrauch nicht-toxischer Alternativmunition und/ oder das Entfernen bleikontaminierter Aufbrüche und Kadaver aus der Landschaft kann dazu einen wichtigen Beitrag leisten.

APPENDIX

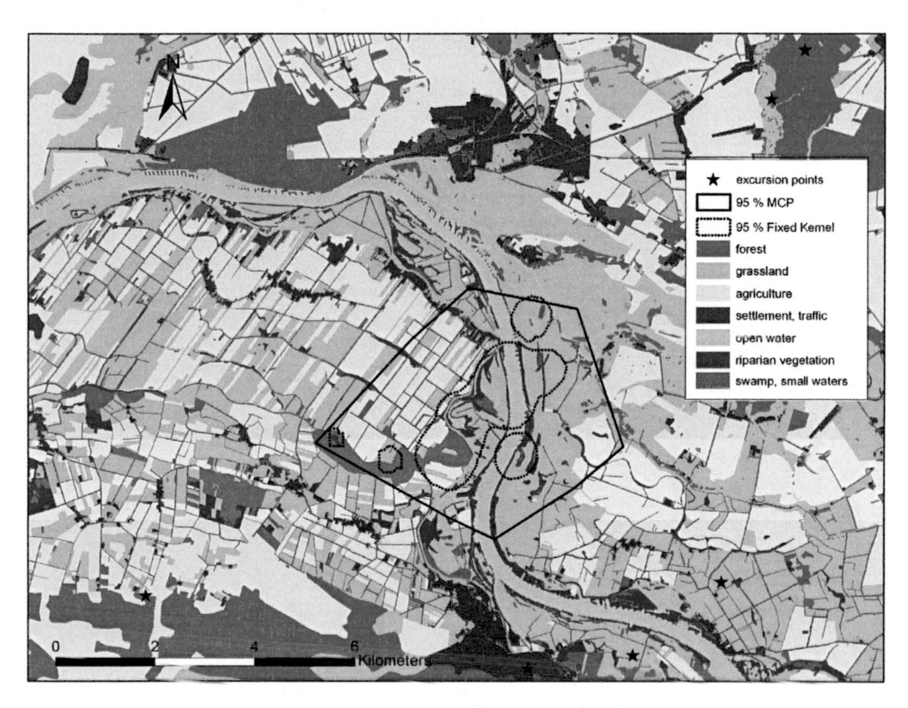

Fig. 1. Home range use of eagle „2355"/ Study site at the Elbe river in the biosphere reserve "*Niedersächsische Elbtalaue*", MCP: minimum convex polygon

Table 1. Individual habitat use at the 'activity range' scale (second order selection), $\hat{w}_{ij}$: selection ratio for the jth animal and the ith resource category, [1]: eagle, values given as $\hat{w}_{ij}$ ± standard error, values > 0: overselected, values < 0: underselected, bold: significant preference, underlined: significant avoidance, **: $p \leq 0.01$, ***: $p \leq 0.001$ (tested by χ^2 log-likelihood statistics, Bonferroni adjusted alpha = 0.0063)

Habitat	$\hat{w}_{ij}$ 655[1]	$\hat{w}_{ij}$ 992[1]	$\hat{w}_{ij}$ 472[1]	$\hat{w}_{ij}$ 964[1]	$\hat{w}_{ij}$ 2907[1]	$\hat{w}_{ij}$ 2905[1]	$\hat{w}_{ij}$ 2355[1]	$\hat{w}_{ij}$ 2906[1]
Forest	<u>0.12 ± 0.08</u>***	**2.25 ± 0.16*** **	**2.25 ± 0.11*** **	**2.12 ± 0.11*** **	**1.92 ± 0.20*** **	**3.06 ± 0.10*** **	<u>0.33 ± 0.10</u>***	**1.85 ± 0.13*** **
Grassland	<u>0.42 ± 0.10</u>***	0.65 ± 0.21	0.63 ± 0.14	0.63 ± 0.14	0.52 ± 0.25	<u>0.30 ± 0.13</u>***	**1.86 ± 0.10*** **	1.34 ± 0.18
Arable land	<u>0.10 ± 0.05</u>***	<u>0.13 ± 0.10</u>***	<u>0.02 ± 0.07</u>***	<u>0.18 ± 0.07</u>***	<u>0.64 ± 0.12</u>**	<u>0.05 ± 0.06</u>***	<u>0.45 ± 0.07</u>***	<u>0.40 ± 0.08</u>***
Settlement	<u>0.05 ± 0.20***</u>	0.16 ± 0.42	<u>0.19 ± 0.29</u>**	0.36 ± 0.28	0.66 ± 0.50	<u>0.29 ± 0.26</u>**	0.35 ± 0.27	0.52 ± 0.36
Traffic	0.22 ± 0.56	0.84 ± 1.14	0.08 ± 0.79	0.64 ± 0.77	1.13 ± 1.34	1.71 ± 0.69	1.17 ± 0.52	1.95 ± 0.96
Open water	**11.93 ± 0.17*** **	**3.82 ± 0.35*** **	**4.80 ± 0.24*** **	**3.99 ± 0.24*** **	1.47 ± 0.47	**2.17 ± 0.23*** **	**6.75 ± 0.40*** **	1.56 ± 0.29
Riparian vegetation	**10.39 ± 0.45*** **	3.12 ± 0.92	2.74 ± 0.64	2.28 ± 0.64	2.92 ± 1.13	**2.94 ± 0.57*** **	**7.18 ± 0.50*** **	1.25 ± 0.78
Swamp/ small waters	0.31 ± 0.34	0.45 ± 0.71	0.44 ± 0.49	1.08 ± 0.49	0.64 ± 0.84	0.30 ± 0.44	**2.89 ± 0.54*** **	0.58 ± 0.60

Table 2. Individual habitat use at the 'within-activity range' scale (third order selection), $\hat{w}_{ij}$: selection ratio for the jth animal and the ith resource category, [1]: eagle, values given as $\hat{w}_{ij}$ ± standard error, values > 0: overselected, values < 0: underselected, bold: significant preference, underlined: significant avoidance, **: $p \leq 0.01$, ***: $p \leq 0.001$ (tested by χ^2 log-likelihood statistics, Bonferroni adjusted alpha = 0.0063)

Habitat	$\hat{w}_{ij}$ 655[1]	$\hat{w}_{ij}$ 992[1]	$\hat{w}_{ij}$ 472[1]	$\hat{w}_{ij}$ 964[1]	$\hat{w}_{ij}$ 2907[1]	$\hat{w}_{ij}$ 2905[1]	$\hat{w}_{ij}$ 2355[1]	$\hat{w}_{ij}$ 2906[1]
Forest	0.10 ± 0.07***	1.04 ± 0.11	0.79 ± 0.04***	**1.79 ± 0.09*** **	**1.39 ± 0.14** **	**1.34 ± 0.06*** **	**2.81 ± 0.18*** **	**1.37 ± 0.11*** **
Grassland	0.42 ± 0.09***	1.43 ± 0.25	1.46 ± 0.18	0.43 ± 0.15***	0.44 ± 0.29	0.45 ± 0.15***	0.76 ± 0.06***	1.19 ± 0.19
Arable land	0.16 ± 0.10***	0.11 ± 0.14***	0.12 ± 0.24***	0.08 ± 0.10***	0.52 ± 0.16**	0.19 ± 0.11***	0.21 ± 0.09***	0.65 ± 0.13
Settlement	0.00 ± 0.00	0.05 ± 0.61	0.02 ± 0.89	0.00 ± 0.38	0.70 ± 0.61	0.08 ± 0.45	0.11 ± 0.33	0.07 ± 0.47
Traffic	0.07 ± 0.65	0.03 ± 0.87	0.04 ± 0.76	0.82 ± 0.82	0.03 ± 1.42	0.29 ± 0.69	0.60 ± 0.40	0.35 ± 1.07
Open water	**1.62 ± 0.07*** **	**3.46 ± 0.31*** **	1.48 ± 0.19	1.16 ± 0.13	1.21 ± 0.37	1.16 ± 0.16	**3.66 ± 0.24*** **	0.46 ± 0.16***
Riparian vegetation	**9.91 ± 0.22*** **	**3.44 ± 0.59*** **	**6.56 ± 0.46*** **	**6.79 ± 0.45*** **	**4.29 ± 0.76*** **	**5.47 ± 0.40*** **	**2.38 ± 0.27*** **	1.87 ± 0.48
Swamp/ small waters	0.31 ± 0.38	0.09 ± 0.76	1.11 ± 0.50	0.65 ± 0.44	1.31 ± 0.94	0.20 ± 0.59	1.21 ± 0.37	**3.07 ± 0.53*** **

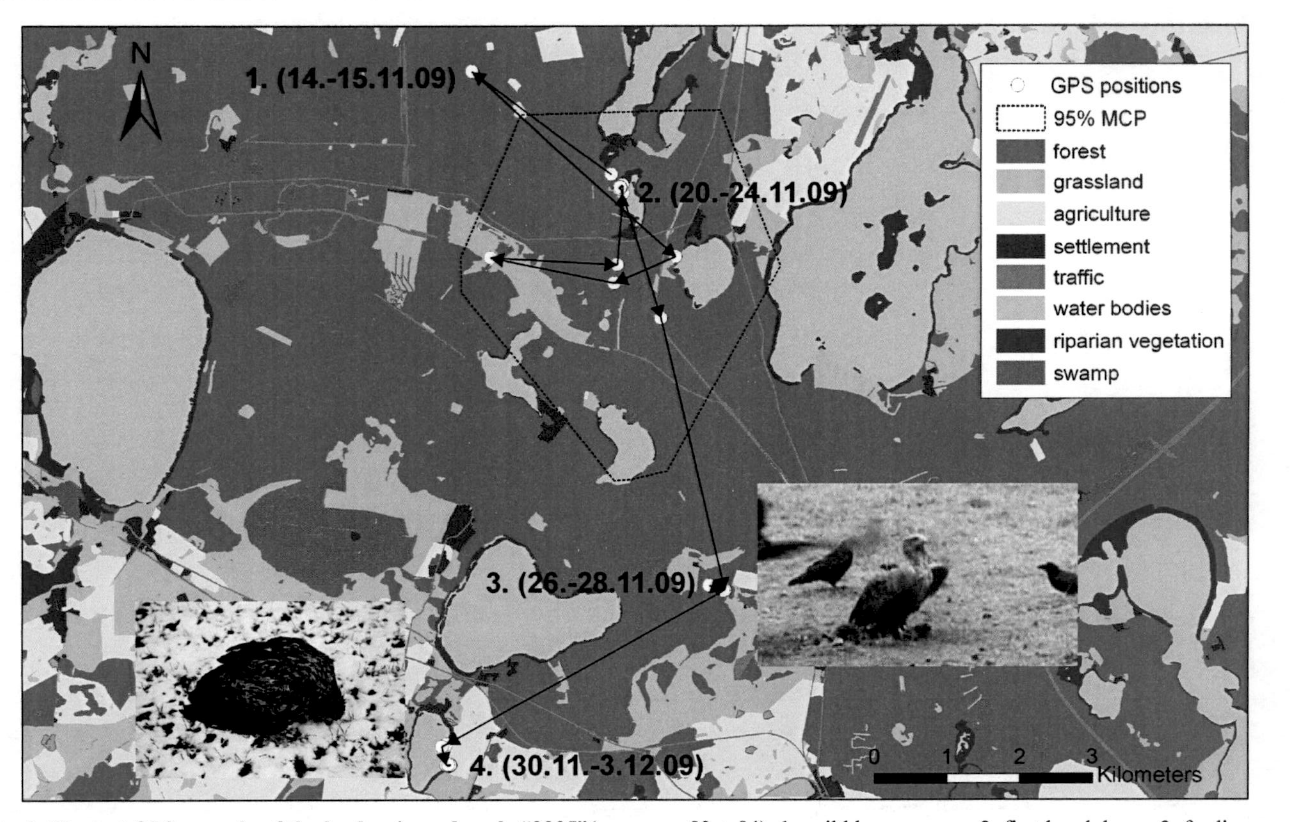

Fig. 2. The last GPS records of the lead-poisoned eagle "2905"(see pages 83 + 84), 1.: wild boar carcass, 2. first breakdown, 3. feeding on a bait at a camera site of M. Nadjafzadeh, 4.: second breakdown and collapse

DANKSAGUNG

Mein herzlicher Dank geht an Dr. Oliver Krone für seine vielfältige Unterstützung und die Möglichkeit, in seinem spannenden Projekt mitarbeiten und ein so hochinteressantes Dissertationsthema bearbeiten zu können. Zudem förderte er diese Dissertation maßgeblich, indem er mir zu Projektbeginn fünf Datensätze von bereits von ihm besenderten Adlern zur Verfügung stellte und mich beim Fangen unterstützte.

Prof. Dr. Heribert Hofer danke ich vielmals für interessante Fachdiskussionen, hilfreiche Anregungen und nützliche Kommentare und Korrekturen, sowie für die Begutachtung dieser Arbeit.

Vielen Dank an Prof. Dr. Silke Kipper für die freundliche Begutachtung dieser Dissertation.

Ich danke dem Bundesministerium für Bildung und Forschung (BMBF) für die Finanzierung dieser Studie, sowie dem Projektträger Jülich für die administrative Betreuung des Gesamtprojekts. Zudem unterstütze mich das IZW finanziell während der Abschlussphase dieser Doktorarbeit.

Ohne die logistische Unterstützung seitens der Mitarbeiter der Naturparkverwaltung Nossentiner/ Schwinzer Heide und des Biosphärenreservat Niedersächsische Elbtalaue wäre diese Arbeit nicht möglich gewesen. Dr. Wolfgang Mewes und Dr. Wolfgang Neubauer haben das Projekt im Naturpark tatkräftig unterstützt, wofür ich ihnen dankbar bin. Ich durfte in der Nossentiner Heide viele unvergessliche Naturimpressionen und Tierbegegnungen erleben und danke allen Beteiligten für die Naturschutzarbeit in der Region.

Ullrich Reeps (Reepsholt-Stiftung) danke ich herzlich für die Unterkunft in seiner Feldstation am wunderschönen Drewitzer See während meiner Feldaufenthalte (siehe Titelblatt). Sein oft rabenschwarzer Humor und Anekdoten aus seinem bewegten Leben haben so machen langen Arbeitstag aufgelockert.

Das Forstamt Sandhof und viele Jäger, Fischer und Landwirte in der Nossentiner Heide haben durch ihre Hilfe bei der Beschaffung der Köder und ihr Einverständnis, private Flächen und Hochstände zum Fangen nutzen zu können, viel zum Projekt beigetragen. Vielen Dank!

Allen Mitgliedern des „Seeadler-Teams“ gilt mein aufrichtiger Dank! Mirjam Nadjafzadeh stand mir immer mit Rat und Tat freundschaftlich zur Seite, verbrachte mit mir große Teile meiner Feldaufenthalte und kennt mich wohl mittlerweile so gut wie nur wenige Menschen. Justine Sulawa munterte mich unzählige Male mit ihrem trockenen Humor auf, war immer offen für allerlei Unternehmungen und beherbergte außerdem meine Mieze während meiner Feldaufenthalte. Danke für Eure Freundschaft, ich werde die schöne Zeit mit Euch nie vergessen, „wir sind für immer die drei Musketiere“! Norbert Kenntners brennender Begeisterung für Habichte konnte ich mich kaum entziehen und für Anna Trinogga konnte man die Hand ins Feuer legen. Danke auch Kirsi Blank und Katrin Totschek für Eure Fröhlichkeit und Euren Teamgeist. Ihr alle habt einen großen Beitrag zum Gelingen dieser Doktorarbeit geleistet.

Vielen Dank auch an Jan Axtner, Kathleen Röllig, Simon Ghanem, Rahel Sollmann, Yvonne Meyer-Lucht, Andreas Wilting, Deike Hesse, Leif Soennichsen, Sarah Benhaiem, Niko Balkenhol und vielen anderen für die schöne Zeit am IZW und in Berlin. An die unzähligen Mittagessensgänge im Tierpark, Grillen im Treptower Park und Kickern im Liberación werde ich mich immer gerne erinnern. Auch für die „seelische und moralische“ Unterstützung meiner Freunde Juliane Holland-Moritz, Dörte Raßbach, Annett Walther, Jakob Maercker und Steffi Knopp möchte ich mich herzlich bedanken.

Niko Balkenhol, Deike Hesse, Stephanie Kramer-Schadt, Rahel Sollmann und Jan Axtner waren so nett, Teile dieser Arbeit zu korrigieren, was sie sehr aufgewertet hat - herzlichen Dank! Danke auch an Stephanie Kramer-Schadt, Nina Klar und Martin Wegmann für GIS-Tipps.

Mein tiefer Dank geht an meine Eltern Agnes Gudrun und Hans-Joachim Scholz, sowie an meine Geschwister Konrad und Ulrike. Ihr habt immer fest an mich geglaubt und in mich vertraut. Dadurch habt ihr mir die Sicherheit vermittelt, meinen Interessen und Träumen zu folgen und meinen Weg durchs Leben zu gehen.

Zuguterletzt möchte ich meine Dankbarkeit den besenderten Seeadlern gegenüber zum Ausdruck bringen, welche die Datengrundlage für diese Dissertation lieferten. Ich hoffe sehr, meine Arbeit konnte einen Beitrag dazu leisten, dass ein so grausamer Tod wie der durch Bleivergiftung zukünftig so viel Seeadlern wie möglich erspart bleibt!